AERODYNAMICS

Gary M. Ullrich

Mark J. Dusenbury

www.kendallhunt.com
Send all inquiries to:
4050 Westmark Drive
Dubuque, IA 52004-1840

ISBN 978-0-7575-9932-3

Printed in the United States of America
10 9 8 7 6 5 4 3 2

TABLE OF CONTENTS

About the Authors

Gary M. Ullrich is an Assistant Professor for the John D. Odegard School of Aerospace Sciences at the University of North Dakota in Grand Forks, North Dakota. Before joining the University of North Dakota, Professor Ullrich was a Test Group Pilot for Flight Safety, Adjunct Assistant Professor for Embry Riddle Aeronautical University, and Instructor/Evaluator Pilot with the United States Air Force. Professor Ullrich is a commercial pilot with multiengine, instrument, and Boeing 720 and 707 type ratings.

Mark Dusenbury is an Assistant Professor for the John D. Odegard School of Aerospace Sciences at the University of North Dakota in Grand Forks, North Dakota. Before coming to the University of North Dakota, Professor Dusenbury was an airline pilot for American Eagle Airlines, and a member of the United States Marine Corps Reserves. Professor Dusenbury holds a commercial pilot certificate with instrument, single, and multi-engine ratings, and is a certified flight instructor for single, multi-engine, and instrument airplane.

Chapter 1

Physical Principles Review

1.0 Algebra

Algebra is the branch of mathematics that uses letters or symbols to represent variables in formulas and equations.

For example, in the equation $D = V \times T$, where Distance = Velocity × Time, the variables are: D, V, and T.

1.1 Equations

Algebraic equations are frequently used in aviation to show the relationship between two or more variables. Equations normally have an equal sign (=) in the expression.

Example: The formula $A = \pi \times r^2$ shows the relationship between the area of a circle (A) and the length of the radius (r) of the circle. The area of a circle is equal to π (3.1416) times the radius squared. Therefore, the larger the radius, the larger the area of the circle.

1.2 Algebraic Rules

When solving for a variable in an equation, you can add, subtract, multiply, or divide the terms in the equation, but you must do the same to both sides of the equal sign.

Example 1: Solve the following equation for the value N.

$$3N = 21$$

To solve for N, divide both sides by 3.

$$3N \div 3 = 21 \div 3$$

$$N = 7$$

Example 2: Solve the following equation for the value N.

$$N + 17 = 59$$

To solve for N, subtract 17 from both sides.

$$N + 17 - 17 = 59 - 17$$

$$N = 42$$

Example 3: Solve the following equation for the value N.

$$N - 22 = 100$$

To solve for N, add 22 to both sides.

$$N - 22 + 22 = 100 + 22$$

$$N = 122$$

1.3 Order of Operation

In algebra, rules have been set for the order in which operations are evaluated. These same universally accepted rules are also used when programming algebraic equations in calculators. When solving the following equation, the **order of operation** is given below:

$$N = (62 - 54)^2 + 6^2 - 4 + 3 \times [8 + (10 \div 2)] + \sqrt{25} + (42 \times 2) \div 4 + 3/4$$

Order of Operation 1. Parentheses. First, do everything in parentheses, ( ). Starting from the innermost parentheses. if the expression has a set of brackets, [ ], treat these exactly like parentheses. If you are working with a fraction, treat the numerator as if it were in parentheses and the denominator as if it were in parentheses, even if there are none shown. From the equation above, completing the calculation in parentheses gives the following:

$$N = (8)^2 + 6^2 - 4 + 3 \times [8 + (5)] + \sqrt{25} + (84) \div 4 + 3/4, \text{ then}$$

$$N = (8)^2 + 6^2 - 4 + 3 \times [13] + \sqrt{25} + 84 \div 4 + 3/4$$

Order of Operation 2. Exponents. Next, clear any exponents. Treat any roots (square roots, cube roots, and so forth) as exponents. Completing the exponents and roots in the equation gives the following:

$$N = 64 + 36 - 4 + 3 \times 13 + 5 + 84 \div 4 + 3/4$$

Order of Operation 3. Multiplication and Division. Evaluate all of the multiplications and divisions from left to right.
Multiply and divide from left to right in one step. A common error is to use two steps for this (that is, to clear all of the multiplication signs and then clear all of the division signs), but this is not the correct method. Treat fractions as division. Completing the multiplication and division in the equation gives the following:

$$N = 64 + 36 - 4 + 39 + 5 + 21 + 3/4$$

Order of Operation 4. Addition and Subtraction. Evaluate the additions and subtractions from left to right. Like above, addition and subtraction are computed left to right in one step. Completing the addition and subtraction in the equation gives the following:

$$N = 161\ 3/4$$

1.4 Order of Operation for Algebraic Equations

1. Parentheses
2. Exponents
3. Multiplication and Division
4. Addition and Subtraction

Use the acronym PEMDAS to remember the order of operation in algebra. **PEMDAS** is an acronym for parentheses, exponents, multiplication, division, addition, and subtraction. To remember it, many use the sentence, "Please Excuse My Dear Aunt Sally." Always remember, however, to multiply/divide or add/subtract in one sweep from left to right, not separately.

1.5 Computing the Area of Two-dimensional Solids

Area is a measurement of the amount of surface of an object. Area is usually expressed in such units as square inches or square centimeters for small surfaces or in square feet or square meters for larger surfaces.

1.6 Rectangle

A **rectangle** is a four-sided figure with opposite sides of equal length and parallel. (Figure 1.1) All of the angles are right angles. A **right angle** is a 90° angle. The rectangle is a very familiar shape in mechanics. The formula for the area of a rectangle is:

$$\text{Area} = \text{Length} \times \text{Width} = L \times W$$

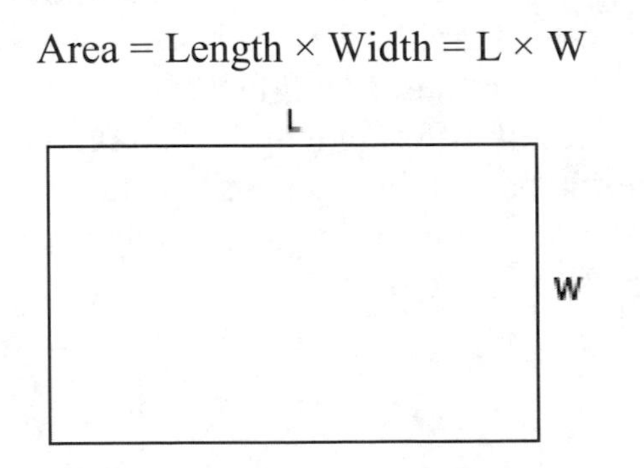

Figure 1.1. Rectangle.

Example: An aircraft floor panel is in the form of a rectangle having a length of 24 inches and a width of 12 inches. What is the area of the panel expressed in square inches? First, determine the known values and substitute them in the formula.

$$A = L \times W = 24 \text{ inches} \times 12 \text{ inches} = 288 \text{ square inches (in}^2\text{)}$$

1.7 Square

A **square** is a four-sided figure with all sides of equal length and parallel. (Figure 1.2) All angles are right angles. The formula for the area of a square is:

$$\text{Area} = \text{Length} \times \text{Width} = L \times W$$

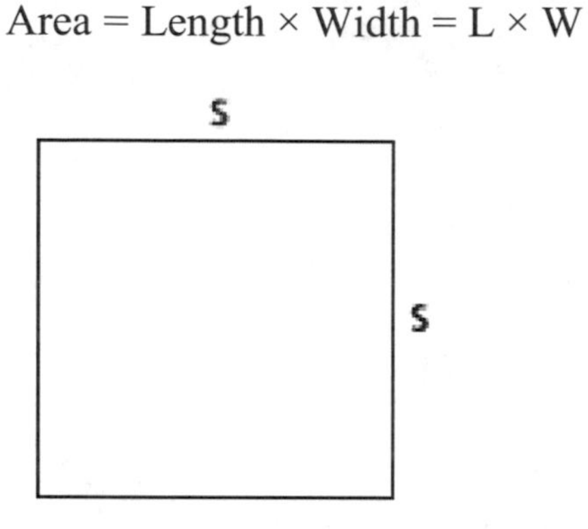

Figure 1.2. Square.

Since the length and the width of a square are the same value, the formula can be written:

$\text{Area} = \text{Side} \times \text{Side} = S^2$

Example: What is the area of a square access plate whose side measures 25 inches? First, determine the known value and substitute it in the formula.

$$A = L \times W = 25 \text{ inches} \times 25 \text{ inches} = 625 \text{ square inches}$$

1.8 Triangle

A **triangle** is a three-sided figure. The sum of the three angles in a triangle is always equal to 180°. Triangles are often classified by their sides. An **equilateral triangle** has three sides of equal length. An **isosceles triangle** has two sides of equal length. A **scalene triangle** has three sides of differing length. Triangles can also be classified by their angles: An **acute triangle** has all three angles less than 90°. A **right triangle** has one right angle (a 90° angle). An **obtuse triangle** has one angle greater than 90°. Each of these types of triangles is shown in Figure 1.3.

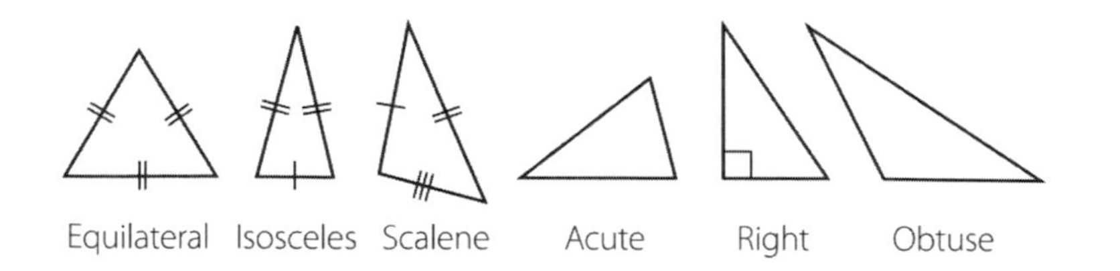

Figure 1.3. Types of Triangles.

The formula for the area of a triangle is:

$$\text{Area} = 1/2 \times (\text{Base} \times \text{Height}) = 1/2 \times (B \times H)$$

Example: Find the area of the obtuse triangle shown in Figure 1.4. First, substitute the known values in the area formula.

$$A = 1/2 \times (B \times H) = 1/2 \times (2'6'' \times 3'2'')$$

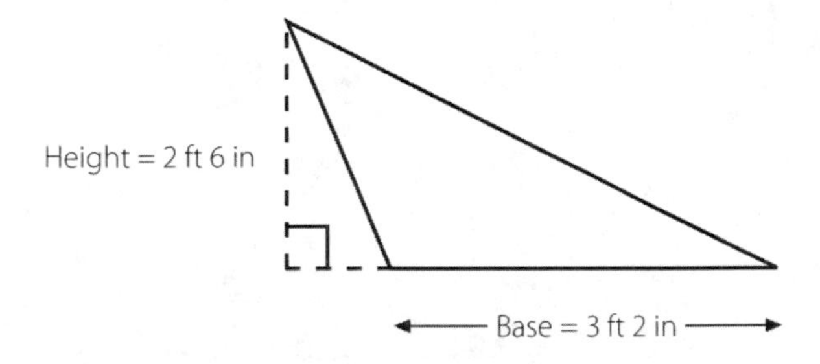

Figure 1.4. Obtuse Triangle.

Next, convert all dimensions to inches:

$$2'6'' = (2 \times 12'') + 6'' = (24 + 6) = 30 \text{ inches}$$
$$3'2'' = (3 \times 12'') + 2'' = (36 + 2) = 38 \text{ inches}$$

Now, solve the formula for the unknown value:

$$A = 1/2 \times (30 \text{ inches} \times 38 \text{ inches}) = 570 \text{ square inches}$$

1.9 Parallelogram

A **parallelogram** is a four-sided figure with two pairs of parallel sides. (Figure 1.5) Parallelograms do not necessarily have four right angles. The formula for the area of a parallelogram is:

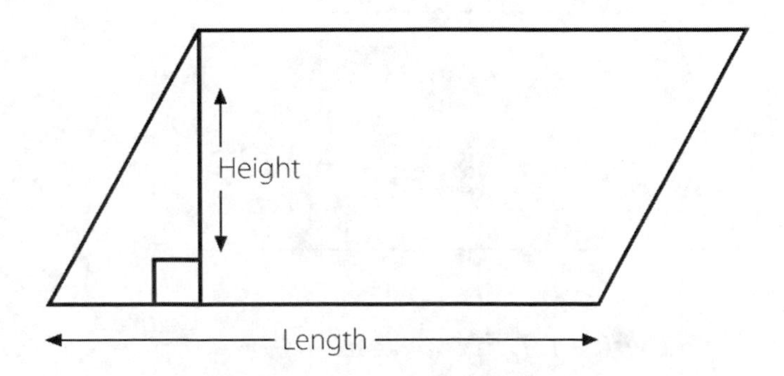

Figure 1.5. Parallelogram.

Area = Length × Height = L × H

1.10 Trapezoid

A **trapezoid** is a four-sided figure with one pair of parallel sides. (Figure 1.6) The formula for the area of a trapezoid is:

$$\text{Area} = 1/2\,(\text{Base}_1 + \text{Base}_2) \times \text{Height}$$

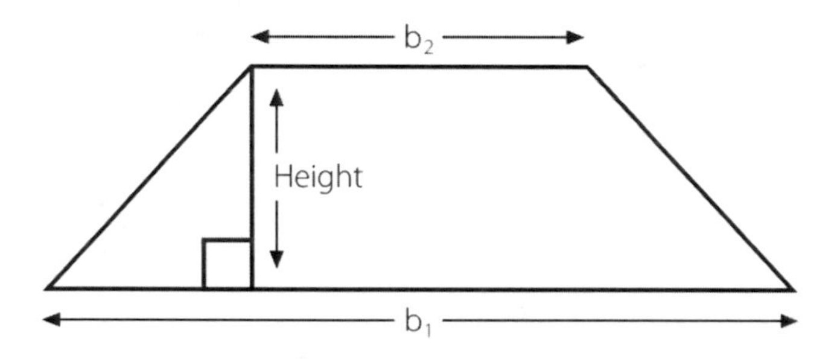

Figure 1.6. Trapezoid.

Example: What is the area of a trapezoid in Figure 1.7 whose bases are 14 inches and 10 inches, and whose height (or altitude) is 6 inches? First, substitute the known values in the formula.

$$A = 1/2\,(b1 + b2) \times H$$

$$A = 1/2\,(14 \text{ inches} + 10 \text{ inches}) \times 6 \text{ inches}$$

$$A = 1/2\,(24 \text{ inches}) \times 6 \text{ inches}$$

$$A = 12 \text{ inches} \times 6 \text{ inches}$$

$$A = 72 \text{ square inches.}$$

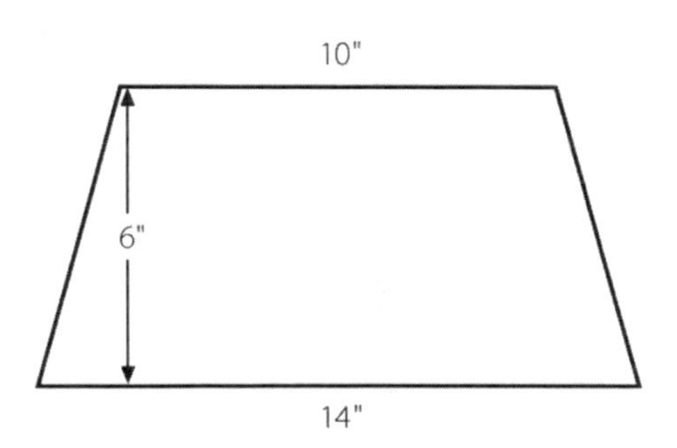

Figure 1.7. Trapezoid.

1.11 Circle

A **circle** is a closed, curved, plane figure. (Figure 1.8) Every point on the circle is an equal distance from the center of the circle. The **diameter** is the distance across the circle (through the center). The **radius** is the distance from the center to the edge of the circle. The diameter is always twice the length of the radius. The **circumference**, or distance around, a circle is equal to the diameter times π. We will assume π to only two significant digits—3.14.

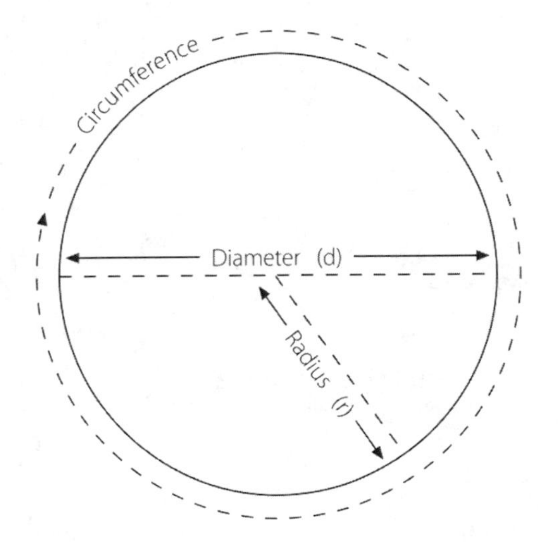

Figure 1.8. Circle.

Circumference = $C = d\,\pi$

The formula for the area of a circle is: Area = $\pi \times$ radius2 = $\pi \times r2$

Example: The **bore**, or "inside diameter," of a certain aircraft engine cylinder is 5 inches. Find the area of the cross section of the cylinder.

First, substitute the known values in the formula: $A = \pi \times r2$.

The diameter is 5 inches, so the radius is 2.5 inches. (diameter = radius × 2)
A = 3.14 × (2.5 inches)2 = 3.14 × 6.25 square inches = 19.635 square inches

1.12 Ellipse

An **ellipse** is a closed, curved, plane figure and is commonly called an oval. (Figure 1.9) In a radial engine, the articulating rods connect to the hub by pins, which travel in the pattern of an ellipse (i.e., an elliptical or orbital path).

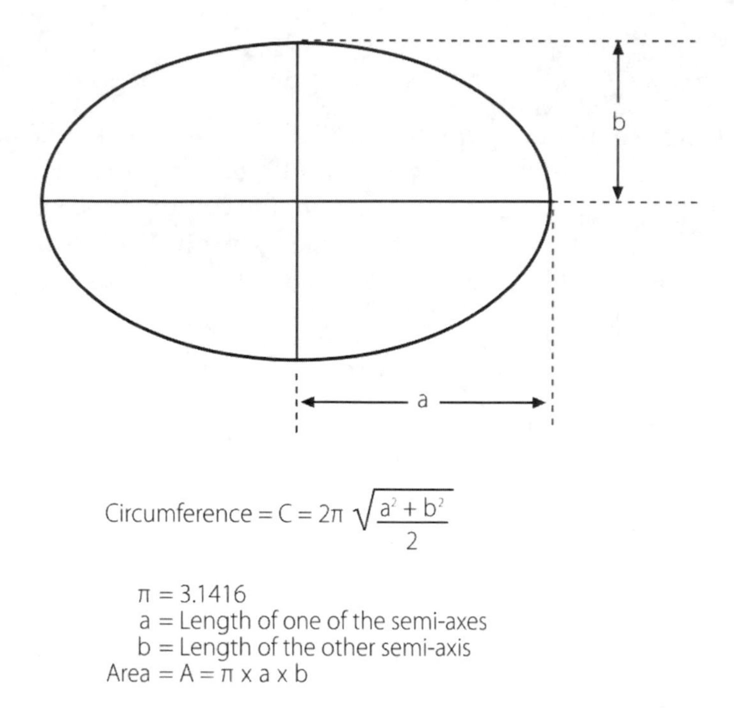

$$\text{Circumference} = C = 2\pi\sqrt{\frac{a^2 + b^2}{2}}$$

$\pi = 3.1416$
a = Length of one of the semi-axes
b = Length of the other semi-axis
Area = $A = \pi \times a \times b$

Figure 1.9. Ellipse.

1.13 Wing Area

To describe the shape of a wing (Figure 1.10), several terms are required. To calculate wing area, it will be necessary to know the meaning of the terms "span" and "chord." The **wingspan**, S, is the length of the wing from wingtip to wingtip. The **chord** is the average width of the wing from leading edge to trailing edge. If the wing is a tapered wing, the average width, known as the mean chord (C), must be known to find the area. The formula for calculating wing area is:

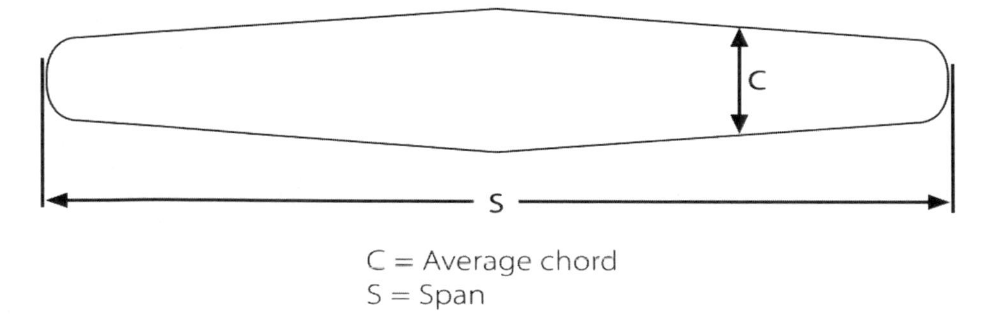

C = Average chord
S = Span

Figure 1.10. Wing Planform.

Area of a wing = Span × Mean Chord. Example: Find the area of a tapered wing whose span is 50 feet and whose mean chord is 6'8". First, substitute the known values in the formula.

$$A = S \times C$$

A= 50 feet × 6 feet 8 inches (Note: 8 inches = 8⁄12 feet = .67 feet)

$$A = 50 \text{ feet} \times 6.67 \text{ feet}$$

$$A = 333.5 \text{ square feet}$$

1.14 Trigonometric Functions

Trigonometry is the study of the relationship between the angles and sides of a triangle. The word trigonometry comes from the Greek *trigonon*, which means three angles, and *metro*, which means measure.

1.15 Right Triangle, Sides and Angles

In Figure 1.11, notice that each angle is labeled with a capital letter. Across from each angle is a corresponding side, each labeled with a lowercase letter. This triangle is a right triangle because angle C is a 90° angle. Side a is opposite from angle A, and is sometimes referred to as the "opposite side." Side b is next to, or adjacent to, angle A and is therefore referred to as the "adjacent side." Side c is always across from the right angle and is referred to as the "**hypotenuse**."

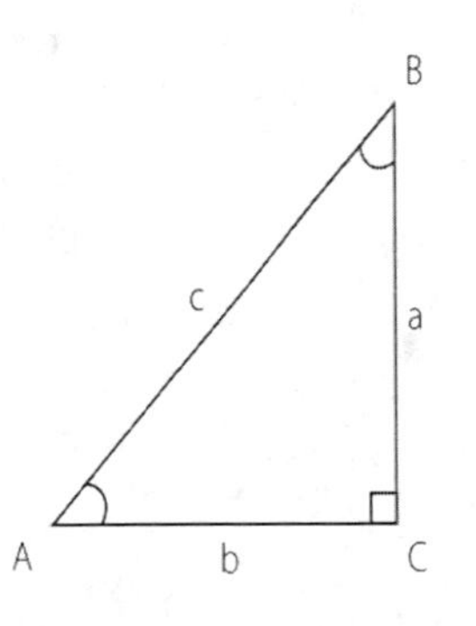

Figure 1.11. Right Triangle.

1.16 Sine, Cosine, and Tangent

The three primary trigonometric functions and their abbreviations are: **sine (sin)**, **cosine (cos)**, and **tangent (tan)**.

These three functions can be found on most scientific calculators. The three trigonometric functions are actually ratios comparing two of the sides of the triangle as follows:

$$\text{Sine (sin) of angle A} = \frac{\text{opposite side (side a)}}{\text{hypotenuse (side c)}}$$

$$\text{Cosine (cos) of angle A} = \frac{\text{adjacent side (side b)}}{\text{hypotenuse (side c)}}$$

$$\text{Tangent (tan) of angle A} = \frac{\text{opposite side (side a)}}{\text{adjacent side (side b)}}$$

Example: Find the sine of a 30° angle.

Calculator Method:

Using a calculator, select the "sin" feature, enter the number 30, and press "enter." The calculator should display the answer as 0.5. This means that when angle A equals 30°, then the ratio of the opposite side (a) to the hypotenuse (c) equals 0.5 to 1, so the hypotenuse is twice as long as the opposite side for a 30° angle. Therefore, sin 30° = 0.5.

1.17 Pythagorean Theorem

The **Pythagorean theorem** is named after the ancient Greek mathematician, Pythagoras (~500 B.C.). This theorem is used to find the third side of any right triangle (Figure 1.12) when two sides are known. The Pythagorean theorem states that $a^2 + b^2 = c^2$. Where c is the hypotenuse of a right triangle, a is one side of the triangle, and b is the other side of the triangle. Example: What is the length of the longest side of a right triangle, given the other sides are 7 inches and 9 inches? The longest side of a right triangle is always side c, the hypotenuse. Use the Pythagorean theorem to solve for the length of side c as follows:

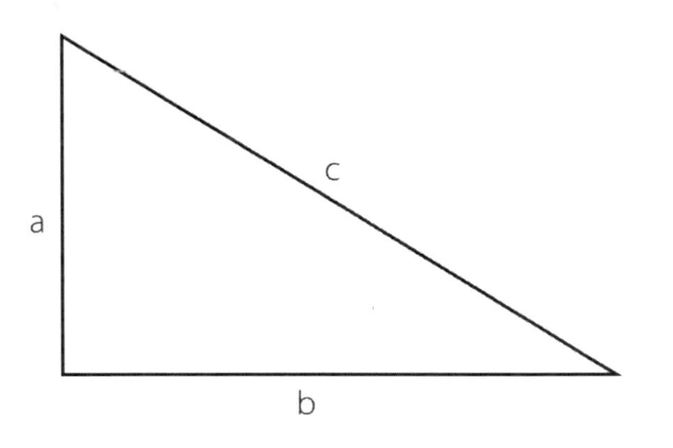

Figure 1.12 Right Triangle.

$$a^2 + b^2 = c^2$$

$$7^2 + 9^2 = c^2$$

$$49 + 81 = c^2$$

$$130 = c^2$$

$$\text{If } c^2 = 130 \text{ then } c = \sqrt{130} = 11.4 \text{ inches}$$

Therefore, side c = 11.4 inches.

Example: The cargo door opening in your airplane is a rectangle that is 5 1⁄2 feet tall by 7 feet wide. A section of square steel plate that is 8 feet wide by 8 feet tall by 1 inch thick must fit inside the airplane. Can the square section of steel plate fit through the cargo door? It is obvious that the square steel plate will not fit horizontally through the cargo door. The steel plate is 8 feet wide and the cargo door is only 7 feet wide. However, if the steel plate is tilted diagonally, will it fit through the cargo door opening? The diagonal distance across the cargo door opening can be calculated using the Pythagorean theorem where "a" is the cargo door height, "b" is the cargo door width, and "c" is the diagonal distance across the cargo door opening.

$$a^2 + b^2 = c^2$$

$$(5.5 \text{ ft})^2 + (7 \text{ ft})^2 = c^2$$

$$30.25 + 49 = c^2$$

$$79.25 = c^2$$

$$c = 8.9 \text{ ft}$$

The diagonal distance across the cargo door opening is 8.9 feet, so the 8-foot wide square steel plate will fit diagonally through the cargo door opening and into the airplane.

1.18 Physics Review

Physical science, or physics as it is most often called, is a very interesting and exciting topic. For the individual who likes technical things and is a hands-on type person, it is an invaluable tool. Physics allows us to explain how engines work, both piston and gas turbine; how airplanes and helicopters fly; and countless other things related to the fields of aviation and aerospace. In addition to allowing us to explain the operation of the things around us, it also allows us to quantify them. For example, through the use of physics we can explain what the concept of thrust means for a jet engine, and then follow it up by mathematically calculating the pounds of thrust being created. **Physics** is the term applied to that area of knowledge regarding the basic and fundamental natures of matter and energy. It does not attempt to determine why matter and energy behave as they do in their relation to physical phenomena, but rather how they behave. The people who maintain and repair aircraft should have a knowledge of basic physics, which is sometimes called the science of matter and energy.

1.19 Matter

Matter is the foundation or the building blocks for any discussion of physics. According to the dictionary, **matter** is what all things are made of; whatever occupies space, has mass, and is perceptible to the senses in some way. According to **the law of conservation**, matter cannot be created or destroyed, but it is possible to change its physical state. When liquid gasoline vaporizes and mixes with air, and then burns, it might seem that this piece of matter has disappeared and no longer exists. Although it no longer exists in the state of liquid gasoline, the matter still exists in the form of the gases given off by the burning fuel.

1.20 Characteristics of Matter

1.20.1 Mass and Weight

Mass is a measure of the quantity of matter in an object. In other words, how many molecules are in the object, or how many atoms, or to be more specific, how many protons, neutrons, and electrons. The mass of an object does not change regardless of where you take it in the universe, and it does not change with a change of state. The only way to change the mass of an object is to add or take away atoms. Mathematically, mass can be stated as follows: The acceleration due to gravity here on earth is 32.2 feet per second per second (32.2 fps/s). An object weighing 32.2 pounds (lb) here on earth is said to have a mass of 1 slug. A **slug** is a quantity of mass that will accelerate at a rate of 1 ft/s^2 when a force of 1 pound is applied. In other words, under standard atmospheric conditions (gravity equal to 32.2) a mass of one slug is equal to 32.2 lb.

Weight is a measure of the pull of gravity acting on the mass of an object. The more mass an object has, the more it will weigh under the earth's force of gravity. Because it is not possible for the mass of an object to go away, the only way for an object to be weightless is for gravity to go away. We view astronauts on the space shuttle and it appears that they are weightless. Even though the shuttle is quite a few miles above the surface of the earth, the force of gravity has not gone away completely, and thus the astronauts are not weightless. The astronauts and the space shuttle are in a state of free fall, so relative to the shuttle the astronauts appear to be weightless. Mathematically, weight can be stated as follows:

Mass = Weight $\div$ Acceleration due to gravity

1.20.2 Attraction

Attraction is the force acting mutually between particles of matter, tending to draw them together. Sir Isaac Newton (1642–1727) called this the "**law of universal gravitation**." He showed how each particle of matter attracts every other particle, how people are bound to the earth, and how the planets are attracted in the solar system.

1.20.3 Porosity

Porosity means having pores or spaces where smaller particles may fit when a mixture takes place. This is sometimes referred to as **granular**—consisting or appearing to consist of small grains or granules.

1.20.4 Impenetrability

Impenetrability means that no two objects can occupy the same place at the same time. Thus, two portions of matter cannot at the same time occupy the same space.

1.20.5 Density

The **density** of a substance is its weight per unit volume. The unit volume selected for use in the English system of measurement is 1 cubic foot (ft^3). In the metric system, it is 1 cubic centimeter (cm^3). Therefore, density is expressed in pounds per cubic foot (lb/ft^3) or in grams per cubic centimeter (g/cm^3).

To find the density of a substance, its weight and volume must be known. Its weight is then divided by its volume to find the weight per unit volume. For example, the liquid which fills a certain container weighs 1,497.6 lb. The container is 4 ft long, 3 ft wide, and 2 ft deep. Its volume is 24 cubic feet or ft^3 (4 ft × 3 ft × 2 ft).

If 24 ft^3 of liquid weighs 1,497.6 lb, then 1 ft^3 weighs 1,497.6 ÷ 24, or 62.4 lb. Therefore, the density of the liquid is 62.4 lb/ft^3. This is the density of water at 4°C (Centigrade) and is usually used as the standard for comparing densities of other substances.

In the metric system, the density of water is 1 g/cm^3. The standard temperature of 4°C is used when measuring the density of liquids and solids. Changes in temperature will not change the weight of a substance, but will change the volume of the substance by expansion or contraction, thus changing its weight per unit volume.

The procedure for finding density applies to all substances; however, it is necessary to consider the pressure when finding the density of gases. Pressure is more critical when measuring the density of gases than it is for other substances. The density of a gas increases in direct proportion to the pressure exerted on it. Standard conditions for the measurement of the densities of gases have been established at 0°C for temperature and a pressure of 76 cm of mercury (Hg). (This is the average pressure of the atmosphere at sea level.) Density is computed based on these conditions for all gases.

1.20.6 Specific Gravity

It is often necessary to compare the density of one substance with that of another. For this purpose, a standard is needed. Water is the standard that physicists have chosen to use when comparing the densities of all liquids and solids. For gases, air is most commonly used. However, hydrogen is sometimes used as a standard for gases. In physics, the word "**specific**" implies a ratio. Thus, **specific gravity** is calculated by comparing the weight of a definite volume of the given substance with the weight of an equal volume of water. The terms “specific weight” or “specific density” are sometimes used to express this ratio. The following formulas are used to find the specific gravity of liquids and solids. The same formulas are used to find the density of gases by substituting air or hydrogen for water. Specific gravity is not expressed in units, but as pure numbers. For example, if a certain hydraulic fluid has a specific gravity of 0.8, then 1 ft^3 of the liquid weighs 0.8 times as much as 1 ft^3 of water: 62.4 times 0.8, or 49.92 lb. Specific gravity and density are independent of the size of the sample under consideration and depend only upon the substance of which it is made. A device called a **hydrometer** is used for measuring specific gravity of liquids. This device consists of a tubular glass float contained in a larger glass tube (Figure 1.13).

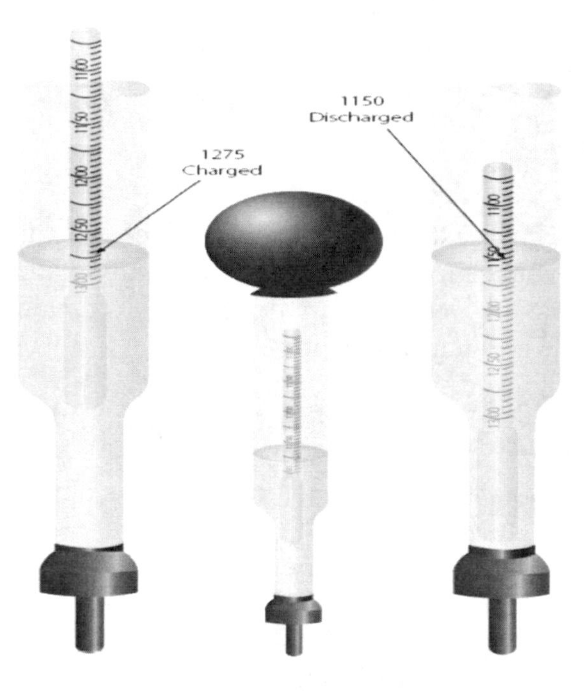

Figure 1.13. Hydrometer.

The larger glass tube provides the container for the liquid. A rubber suction bulb draws the liquid up into the container. There must be enough liquid to raise the float and prevent it from touching the bottom. The float is weighted and has a vertically graduated scale. To determine specific gravity, the scale is read at the surface of the liquid in which the float is immersed. An indication of 1000 is read when the float is immersed in pure water. When immersed in a liquid of greater density, the float rises, indicating a greater specific gravity. For liquids of lesser density the float sinks, indicating a lower specific gravity. An example of the use of the hydrometer is to determine the specific gravity of the electrolyte (battery).

1.21 Simple Machines

1.21.1 What is a Machine?

A **machine** is any device with which work may be accomplished. In application, machines can be used for any of the following purposes, or combinations of these purposes.

- Machines are used to transform energy, as in the case of a generator transforming mechanical energy into electrical energy.

- Machines are used to transfer energy from one place to another, as in the examples of the connecting rods, crankshaft, and reduction gears transferring energy from an aircraft's engine to its propeller.

- Machines are used to multiply force; for example, a system of pulleys may be used to lift a heavy load. The pulley system enables the load to be raised by exerting a force that is smaller than the weight of the load.

- Machines can be used to multiply speed. A good example is the bicycle, by which speed can be gained by exerting a greater force via the gear.

- Machines can be used to change the direction of a force. An example of this use is the flag hoist. A downward force on one side of the rope exerts an upward force on the other side, raising the flag toward the top of the pole.

There are only six simple machines. They are the **lever**, the **pulley**, the **wheel and axle**, the **inclined plane**, the **screw**, and the **gear**.

Physicists, however, recognize only two basic principles in machines: the lever and the inclined plane. The pulley (block and tackle), the wheel and axle, and gears operate on the machine principle of the lever. The wedge and the screw use the principle of the inclined plane.

An understanding of the principles of simple machines provides a necessary foundation for the study of compound machines, which are combinations of two or more simple machines.

1.21.2 Mechanical Advantages of Machines

As identified in statements 3 and 4 under simple machines, a machine can be used to multiply force or to multiply speed. It cannot, however, multiply force and speed at the same time. In order to gain one, it must lose the other. To do otherwise would mean the machine has more power going out than coming in, and that is not possible.

In reference to machines, **mechanical advantage** is a comparison of the output force to the input force, or the output distance to the input distance. If there is a mechanical advantage in terms of force, there will be a fractional disadvantage in terms of distance. The following formulas can be used to calculate mechanical advantage.

Mechanical Advantage = Force Out ÷ Force In

Or

Mechanical Advantage = Distance Out ÷ Distance In

1.21.3 The Lever

The simplest machine, and perhaps the most familiar one, is the lever. A seesaw is a familiar example of a lever, with two people sitting on either end of a board and a pivoting point in the middle. There are three basic parts in all levers. They are the **fulcrum** "F," a force or **effort** "E," and a **resistance** "R." Shown in Figure 1.14 are the pivot point "F" (fulcrum), the effort "E" which is applied at a distance "L" from the fulcrum, and a resistance "R" which acts at a distance "l" from the fulcrum. Distances "L" and "l" are the lever arms.

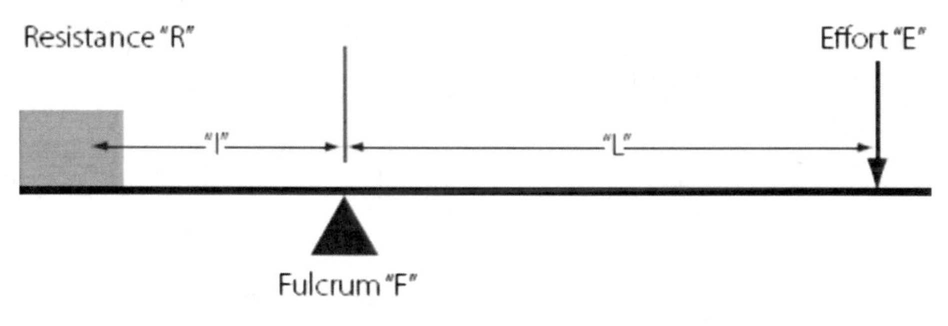

Figure 1.14. First Class Lever.

Torque is very much involved in the operation of a lever. When a person sits on one end of a seesaw, that person applies a downward force in pounds which acts along the distance to the center of the seesaw. This combination of force and distance creates torque, which tries to cause rotation.

1.21.4 First Class Lever

In the first class lever (Figure 1.14), the fulcrum is located between the effort and the resistance. As mentioned earlier, the seesaw is a good example of a lever, and it happens to be a first class lever. The amount of weight and the distance from the fulcrum can be varied to suit the need. Increasing the distance from the applied effort to the fulcrum, compared to the distance from the fulcrum to the weight being moved, increases the advantage provided by the lever. Crowbars, shears, and pliers are common examples of this class of lever. The proper balance of an airplane is also a good example, with the center of lift on the wing being the pivot point (fulcrum) and the weight fore and aft of this point being the effort and the resistance.

When calculating how much effort is required to lift a specific weight, or how much weight can be lifted by a specific effort, the following formula can be used.

$$\text{Effort (E)} \times \text{Effort Arm (L)} = \text{Resistance (R)} \times \text{Resistance Arm (l)}$$

What this formula really shows is the **input torque** (effort times effort arm) equals the **output torque** (resistance times resistance arm). This formula and concept apply to all three classes of levers, and to all simple machines in general.

Example: A first class lever is to be used to lift a 500-lb weight. The distance from the weight to the fulcrum is 12 inches and from the fulcrum to the applied effort is 60 inches. How much force is required to lift the weight?

$$\text{Effort (E)} \times \text{Effort Arm (L)} = \text{Resistance (R)} \times \text{Resistance Arm (l)}$$

$$E \times 60 \text{ in} = 500 \text{ lb} \times 12 \text{ in}$$

$$E = 500 \text{ lb} \times 12 \text{ in} \div 60 \text{ in}$$

$$E = 100 \text{ lb}$$

The mechanical advantage of the lever in this example would be:

$$\text{Mechanical Advantage} = \text{Force Out} \div \text{Force In}$$

$$= 500 \text{ lb} \div 100 \text{ lb}$$

$$= 5, \text{ or } 5 \text{ to } 1$$

An interesting thing to note with this example lever is if the applied effort moved down 10 inches, the weight on the other end would move up only 2 inches. The weight being lifted would only move one-fifth as far. The reason for this is the concept of work. Because a lever cannot have more work output than input, if it allows you to lift five times more weight, you will move it only one-fifth as far as you move the effort.

1.21.5 Second Class Lever

The second class lever (Figure 1.15) has the fulcrum at one end and the effort is applied at the other end. The resistance is somewhere between these points. A wheelbarrow is a good example of a second class lever, with the wheel at one end being the fulcrum, the handles at the opposite end being the applied effort, and the bucket in the middle being where the weight or resistance is placed.

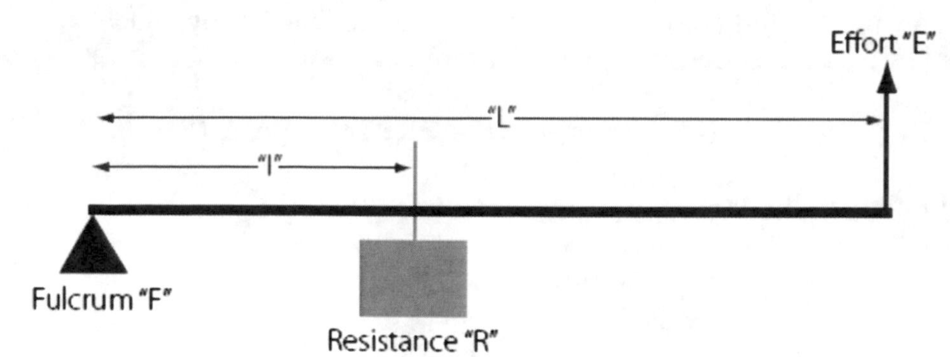

Figure 1.15. Second Class Lever.

Both first and second class levers are commonly used to help in overcoming big resistances with a relatively small effort. The first class lever, however, is more versatile. Depending on how close or how far away the weight is placed from the fulcrum, the first class lever can be made to gain force or gain distance, but not both at the same time. The second class lever can only be made to gain force.

Example: The distance from the center of the wheel to the handles on a wheelbarrow is 60 inches. The weight in the bucket is 18 inches from the center of the wheel. If 300 lb is placed in the bucket, how much force must be applied at the handles to lift the wheelbarrow?

$$\text{Effort (E)} \times \text{Effort Arm (L)} = \text{Resistance (R)} \times \text{Resistance Arm (l)}$$

$$E \times 60 \text{ inches} = 300 \text{ lb} \times 18 \text{ in}$$

$$E = 300 \text{ lb} \times 18 \text{ in} \div 60 \text{ in}$$

$$E = 90 \text{ lb}$$

The mechanical advantage of the lever in this example would be:

$$\text{Mechanical Advantage} = \text{Force Out} \div \text{Force In}$$

$$= 300 \text{ lb} \div 90 \text{ lb}$$

$$= 3.33, \text{ or } 3.33 \text{ to } 1$$

1.21.6 Third Class Lever

There are occasions when it is desirable to speed up the movement of the resistance even though a large amount of effort must be used. Levers that help accomplish this are third class levers. As shown in Figure 1.16, the fulcrum is at one end of the lever and the weight or resistance to be overcome is at the other end, with the effort applied at some point between. Third class levers are easily recognized because the effort is applied between the fulcrum and the resistance. The retractable main landing gear on an airplane is a good example of a third class lever. The top of the landing gear, where it attaches to the airplane, is the pivot point. The wheel and brake assembly at the bottom of the landing gear is the resistance. The hydraulic actuator that makes the gear retract is attached somewhere in the middle, and that is the applied effort.

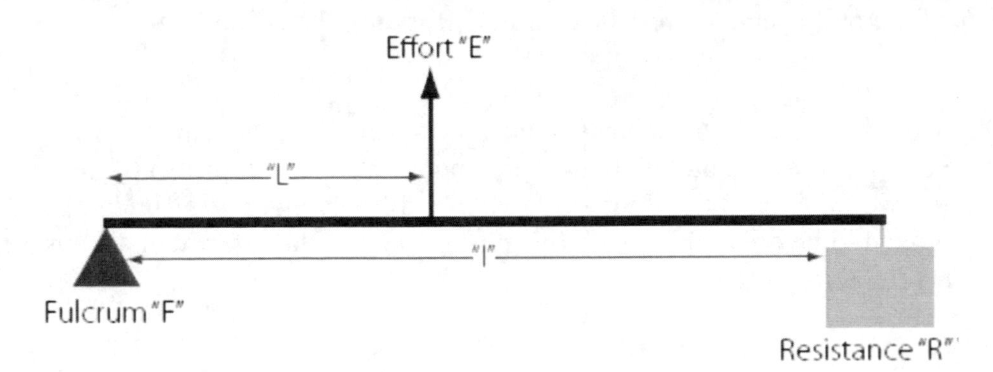

Figure 1.16. Third Class Lever.

1.22 The Pulley

Pulleys are simple machines in the form of a wheel mounted on a fixed axis and supported by a frame. The wheel, or disk, is normally grooved to accommodate a rope. The wheel is sometimes referred to as a "sheave" (sometimes "sheaf"). The frame that supports the wheel is called a block. A block and tackle consists of a pair of blocks. Each block contains one or more pulleys and a rope connecting the pulley(s) of each block.

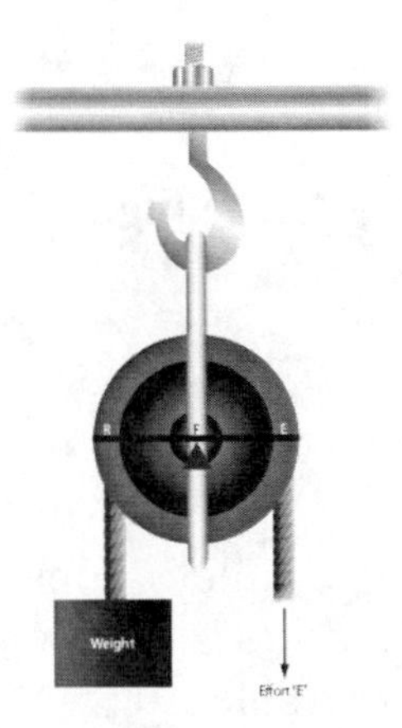

Figure 1.17. Single Fixed Pulley.

1.22.1 Single Fixed Pulley

A single fixed pulley is really a first class lever with equal arms. In Figure 1.17, the arm from point "R" to point "F" is equal to the arm from point "F" to point "E" (both distances being equal to the radius of the pulley). When a first class lever has equal arms, the mechanical advantage is one. Thus, the force of the pull on the rope must be equal to the weight of the object being lifted. The only advantage of a single fixed pulley is to change the direction of the force, or pull on the rope.

1.22.2 Single Movable Pulley

A single pulley can be used to magnify the force exerted. In Figure 1.18, the pulley is movable, and both ropes extending up from the pulley are sharing in the support of the weight. This single movable pulley acts like a second class lever, with the effort arm (EF) being the diameter of the pulley and the resistance arm (FR) being the radius of the pulley.

This type of pulley would have a mechanical advantage of two because the diameter of the pulley is double the radius of the pulley. In use, if someone pulled in four feet of the effort rope, the weight would rise off the floor only two feet. If the weight was 100 lb, the effort applied would need to be only 50 lb. With this type of pulley, the effort will always be one-half of the weight being lifted.

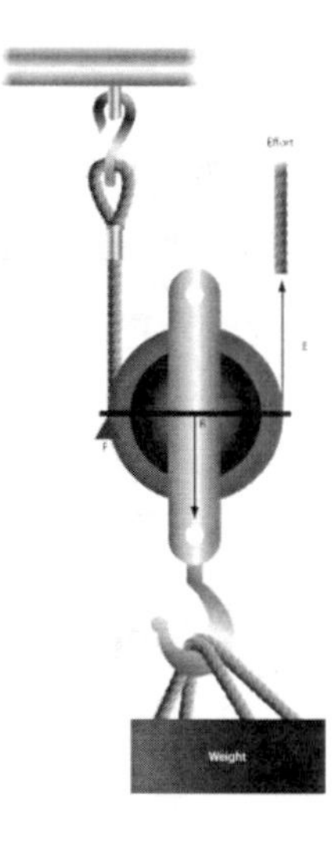

Figure 1.18. Single Movable Pulley.

1.22.2 Block and Tackle

A block and tackle is made up of multiple pulleys, some of them fixed and some movable. In Figure 1.19, the block and tackle is made up of four pulleys, the top two being fixed and the bottom two being movable. Viewing the figure from right to left, notice there are four ropes supporting the weight and a fifth rope where the effort is applied.

The number of weight-supporting ropes determines the mechanical advantage of a block and tackle, so in this case the mechanical advantage is four. If the weight was 200 lb, it would require a 50 lb effort to lift it.

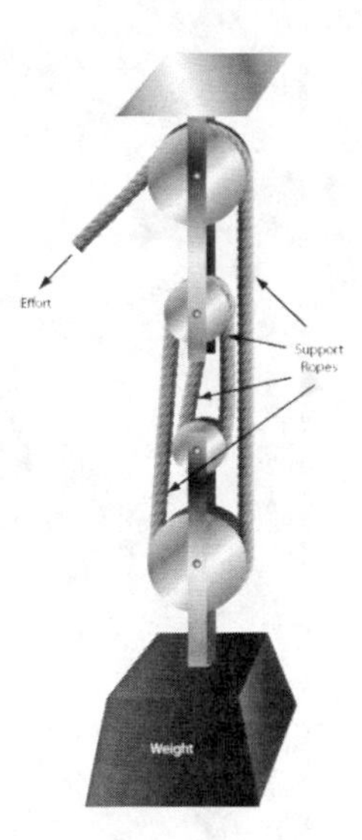

Figure 1.19. Block and Tackle Pulley.

1.23 Motion

The study of the relationship between the motion of bodies or objects and the forces acting on them is often called the study of "force and motion." In a more specific sense, the relationships among velocity, acceleration, and distance is known as **kinematics**.

1.23.1 Uniform Motion

Motion may be defined as a continuing change of position or place, or as the process in which a body undergoes displacement. When an object is at different points in space at different times, that object is said to be in motion, and if the distance the object moves remains the same for a given period of time, the motion may be described as **uniform**. Thus, an object in uniform motion always has a constant speed.

1.23.2 Speed and Velocity

In everyday conversation, speed and velocity are often used as if they mean the same thing. In physics, however, they have definite and distinct meanings. **Speed** refers to how fast an object is moving, or how far the object will travel in a specific time. The speed of an object tells nothing about the direction an object is moving. For example, if the information is supplied that an airplane leaves New York City and travels 8 hours at a speed of 150 mph, this information tells nothing about the direction in which the airplane is moving. At the end of eight hours, it might be in Kansas City, or if it traveled in a circular route, it could be back in New York City. Velocity is that quantity in physics which denotes both the speed of an object and the direction in which the object moves.

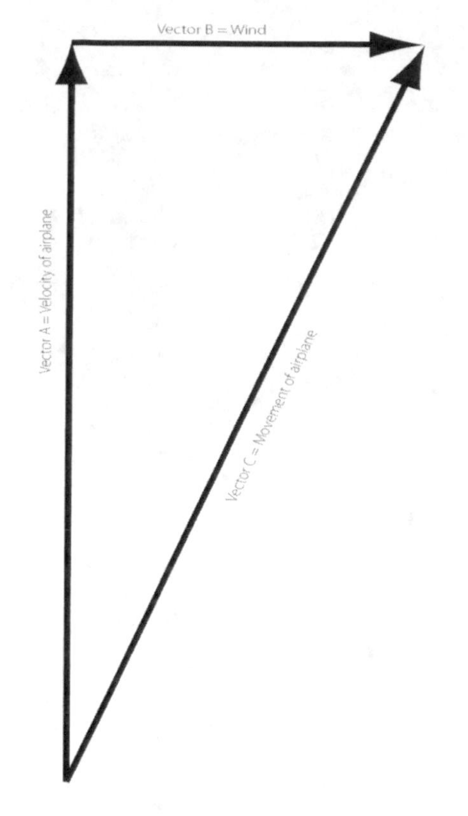

Figure 1.20. Vector Addition.

Velocity can be defined as the rate of motion in a particular direction. Velocity is also described as being a vector quantity, a **vector** being a line of specific length, having an arrow on one end or the other. The length of the line indicates the number value and the arrow indicates the direction in which that number is acting. Two velocity vectors, such as one representing the velocity of an airplane and one representing the velocity of the wind, can be added together in what is called **vector analysis**. Figure 1.20 demonstrates this, with vectors "A" and "B" representing the velocity of the airplane and the wind, and vector "C" being the resultant. With no wind, the speed and direction of the airplane would be that shown by vector "A." When accounting for the wind direction and speed, the airplane ends up flying at the speed and direction shown by vector "C."

Imagine that an airplane is flying in a circular pattern at a constant speed. Because of the circular pattern, the airplane is constantly changing direction, which means the airplane is constantly changing velocity. The reason for this is the fact that velocity includes direction. To calculate the speed of an object, the distance it travels is divided by the elapsed time. If the distance is measured in miles and the time in hours, the units of speed will be miles per hour (mph). If the distance is measured in feet and the time in seconds, the units of speed will be feet per second (fps). *To convert mph to fps, multiply by 1.467.* Velocity is calculated the same way, the only difference being it must be recalculated every time the direction changes. **Acceleration** is defined as the rate of change of velocity.

If the velocity of an object is increased from 20 mph to 30 mph, the object has been accelerated. If the increase in velocity is 10 mph in 5 seconds, the rate of change in velocity is 10 mph in 5 seconds, or 2 mph per second. If this were multiplied by 1.467, it could also be expressed as an acceleration of 2.93 feet per second per second (fps/s). By comparison, the acceleration due to gravity is 32.2 fps/s. To calculate acceleration, the following formula is used.

$$\text{Acceleration (A)} = \frac{\text{Velocity Final (Vf)} - \text{Velocity Initial (Vi)}}{\text{Time (t)}}$$

Example: An Air Force F-15 fighter is cruising at 400 knots. The pilot advances the throttles to full afterburner and accelerates to 800 knots in 20 seconds. What is the average acceleration in mph/s and fps/s?

$$A = \frac{Vf - Vi}{T}$$

$$A = \frac{800 - 400}{20}$$

A = 20 knots (nautical miles per hour) per second or by multiplying by 1.467, 29.34 fps/s

1.24 Newton's Laws of Motion

1.24.1 Newton's First Law of Motion

When a magician snatches a tablecloth from a table and leaves a full setting of dishes undisturbed, he is not displaying a mystic art; he is demonstrating the principle of **inertia**. Inertia is responsible for the discomfort felt when an airplane is brought to a sudden halt in the parking area and the passengers are thrown forward in their seats. Inertia is a property of matter. This property of matter is described by Newton's **first law of motion**, which states:

Objects at rest tend to remain at rest and objects in motion tend to remain in motion at the same speed and in the same direction, unless acted on by an external force.

1.24.2 Newton's Second Law of Motion

Bodies in motion have the property called **momentum**. A body that has great momentum has a strong tendency to remain in motion and is therefore hard to stop. For example, a train moving at even low velocity is difficult to stop because of its large mass. Newton's **second law of motion** applies to this property. It states:

When a force acts upon a body, the momentum of that body is changed. The rate of change of momentum is proportional to the applied force. Based on Newton's second law, the formula for calculating **thrust** is derived, which states that force equals mass times acceleration (F = MA).

Earlier in this chapter, it was determined that mass equals weight divided by gravity, and acceleration equals velocity final minus velocity initial divided by time. Putting all these concepts together, the formula for thrust is:

$$\text{Force} = \frac{\text{Weight (Velocity final} - \text{Velocity initial)}}{\text{Gravity (Time)}}$$

$$\text{Force} = \frac{W\,(Vf - Vi)}{Gt}$$

Example: A turbojet engine is moving 150 lb of air per second through the engine. The air enters going 100 fps and leaves going 1,200 fps. How much thrust, in pounds, is the engine creating?

$$F = \frac{W\,(Vf - Vi)}{Gt}$$

$$F = \frac{150\,(1200 - 100)}{32.2\,(1)}$$

$$F = 5{,}124 \text{ lb of thrust}$$

1.24.3 Newton's Third Law of Motion

Newton's **third law of motion** is often called the law of action and reaction. It states that for every action there is an equal and opposite reaction. This means that if a force is applied to an object, the object will supply a resistive force exactly equal to and in the opposite direction of the force applied. It is easy to see how this might apply to objects at rest. For example, as a man stands on the floor, the floor exerts a force against his feet exactly equal to his weight. But this law is also applicable when a force is applied to an object in motion.

Forces always occur in pairs. The term "**acting force**" means the force one body exerts on a second body, and "**reacting force**" means the force the second body exerts on the first.

When an aircraft propeller pushes a stream of air backward with a force of 500 lb, the air pushes the blades forward with a force of 500 lb. This forward force causes the aircraft to move forward. A turbofan engine exerts a force on the air entering the inlet duct, causing it to accelerate out the fan duct and the tailpipe. The air accelerating to the rear is the action, and the force inside the engine that makes it happen is the reaction, also called thrust.

1.25 Circular Motion

Circular motion is the motion of an object along a curved path that has a constant radius. For example, if one end of a string is tied to an object and the other end is held in the hand, the object can be swung in a circle. The object is constantly deflected from a straight (linear) path by the pull exerted on the string, as shown in Figure 1.21. When the weight is at point A, due to inertia it wants to keep moving in a straight line and end up at point B. Because of the force being exerted on the string, it is forced to move in a circular path and end up at point C.

The string exerts a **centripetal force** on the object, and the object exerts an equal but opposite force on the string, obeying Newton's third law of motion. The force that is equal to centripetal force, but acting in an opposite direction, is called **centrifugal force**. Centripetal force is always directly proportional to the mass of the object in circular motion. Thus, if the mass of the object in Figure 1.21 is doubled, the pull on the string must be doubled to keep the object in its circular path, provided the speed of the object remains constant.

Centripetal force is inversely proportional to the radius of the circle in which an object travels. If the string in Figure 1.21 is shortened and the speed remains constant, the pull on the string must be increased since the radius is decreased, and the string must pull the object from its linear path more rapidly. Using the same reasoning, the pull on the string must be increased if the object is swung more rapidly in its orbit. Centripetal force is thus directly proportional to the square of the velocity of the object. The formula for centripetal force is:

$$\text{Centripetal Force} = \frac{\text{Mass (Velocity}^2\text{)}}{\text{Radius}}$$

For the formula above, mass would typically be converted to weight divided by gravity, velocity would be in feet per second, and the radius would be in feet.

Example: What would the centripetal force be if a 10 lb weight was moving in a 3-ft radius circular path at a velocity of 500 fps?

$$\text{Centripetal Force} = \text{Mass (Velocity}^2\text{)} \div \text{Radius}$$

$$\text{Centripetal Force} = 10\ (500^2) \div 32.2\ (3)$$

$$= 25{,}880 \text{ lb}$$

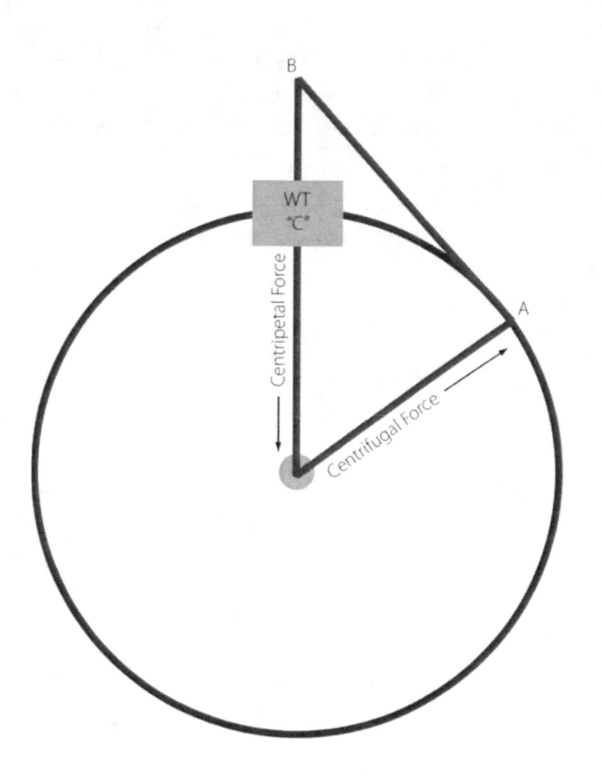

Figure 1.21. Circular Motion.

In the conditions identified in the example, the object acts like it weighs 2,588 times more than it actually does. It can also be said that the object is experiencing 2,588 **Gs** (force of gravity). The fan blades in a large turbofan engine, when the engine is operating at maximum rpm, are experiencing many thousands of Gs for the same reason.

1.26 Heat and Energy

Heat is a form of energy. It is produced only by the conversion of one of the other forms of energy. **Heat** may also be defined as the total kinetic energy of the molecules of any substance. Some forms of energy which can be converted into heat energy are as follows:

- ***Mechanical Energy***. This includes all methods of producing increased motion of molecules such as friction, impact of bodies, or compression of gases.

- ***Electrical Energy***. Electrical energy is converted to heat energy when an electric current flows through any form of resistance such as an electric iron, electric light, or an electric blanket.

- ***Chemical Energy***. Most forms of chemical reaction convert stored potential energy into heat. Some examples are the explosive effects of gunpowder; the burning of aviation gas, oil, or wood; and the combining of oxygen and grease.

• ***Radiant Energy***. Electromagnetic waves of certain frequencies produce heat when they are absorbed by the bodies they strike, such as x-rays, light rays, and infrared rays.

• ***Nuclear Energy***. Energy stored in the nucleus of atoms is released during the process of nuclear fission in a nuclear reactor or atomic explosion.

• ***Thermal Energy***. All heat energy can be directly or indirectly traced to the nuclear reactions occurring in the sun. When a gas is compressed, work is done, and the gas becomes warm or hot. Conversely, when a gas under high pressure is allowed to expand, the expanding gas becomes cool. In the first case, work was converted into energy in the form of heat; in the second case heat energy was expended. Since heat is given off or absorbed, there must be a relationship between heat energy and work. Also, when two surfaces are rubbed together, the friction develops heat. However, work was required to cause the heat, and by experimentation, it has been shown that the work required and the amount of heat produced by friction are proportional. Thus, heat can be regarded as a form of energy. According to this theory of heat as a form of energy, the molecules, atoms, and electrons in all bodies are in a continual state of motion. In a hot body, these small particles possess relatively large amounts of kinetic energy, but in cooler bodies they have less. Because the small particles are given motion, and hence kinetic energy, work must be done to slide one body over the other. Mechanical energy apparently is transformed, and what we know as heat is really kinetic energy of the small molecular subdivisions of matter.

1.26.1 Heat Energy Units

Two different units are used to express quantities of heat energy. They are the calorie and the BTU. One **calorie** is equal to the amount of heat required to change the temperature of 1 gram of water 1 degree Centigrade. This term "calorie" (spelled with a lower case c) is 1/1,000 of the **Calorie** (spelled with a capital C) used in the measurement of the heat energy in foods. One **BTU** is defined as the amount of heat required to change the temperature of 1 lb of water 1 degree Fahrenheit (1°F). The calorie and the gram are seldom used in discussing aviation maintenance. The BTU, however, is commonly referred to in discussions of engine thermal efficiencies and the heat content of aviation fuel. A device known as the **calorimeter** is used to measure quantities of heat energy. For example, it may be used to determine the quantity of heat energy available in 1 pound of aviation gasoline. A given weight of the fuel is burned in the calorimeter, and the heat energy is absorbed by a large quantity of water. From the weight of the water and the increase in its temperature, it is possible to compute the heat yield of the fuel. A definite relationship exists between heat and mechanical energy. This relationship has been established and verified by many experiments which show that:

One BTU of heat energy = 778 ft-lb of work

As will be discussed later under the topic "Potential Energy," one pound of aviation gasoline contains 18,900 BTU of heat energy.

Since each BTU is capable of 778 ft-lb of work, 1 lb of aviation gasoline is capable of 14,704,200 ft-lb of work.

1.26.2 Heat Energy and Thermal Efficiency

Thermal efficiency is the relationship between the potential for power contained in a specific heat source, and how much usable power is actually created when that heat source is used. The formula for calculating thermal efficiency is:

Thermal Efficiency = Horsepower (hp) Produced ÷ Potential Horsepower in Fuel

For example, consider the piston engine used in a small general aviation airplane, which typically consumes 0.5 lb of fuel per hour for each horsepower it creates. Imagine that the engine is creating 200 hp. If we multiply 0.5 by the horsepower of 200, we find the engine is consuming 100 lb of fuel per hour, or 1.67 lb per minute.

One **horsepower** (hp) equates to be 33,000 ft-lb of work per minute.

The potential horsepower in the fuel burned for this example engine would be:

$$\text{hp} = \frac{1.67 \text{ lb per minute} \times 18{,}900 \text{ BTU per lb} \times 778 \text{ ft lb per BTU}}{33{,}000 \text{ ft-lb/min}}$$

$$\text{hp} = 744$$

The example engine is burning enough fuel that it has the potential to create 744 horsepower, but it is only creating 200. The thermal efficiency of the engine would be:

$$\text{Thermal Efficiency} = \text{hp Produced} \div \text{hp in Fuel}$$

$$= 200 \div 744$$

$$= .2688 \text{ or } 26.88\%$$

More than 70 percent of the energy in the fuel is not being used to create usable horsepower. The wasted energy is in the form of friction and heat. A tremendous amount of heat is given up to the atmosphere and not used inside the engine to create power.

Note: Figures in Chapter 1 were taken from the Aviation Maintenance Technician Handbook, FAA-8083-30.

Chapter 2

The Atmosphere

The **atmosphere** is the gaseous envelope covering the earth that is held in place by gravity. Comparing the earth to a baseball, the atmosphere, in perspective, would be about as thick as the baseball's cover. This envelope of air rotates with the earth. The atmosphere also has motions relative to the earth's surface called **circulations**. Circulations are caused primarily by the large temperature difference between the tropics and the polar regions, with other significant factors such as the uneven heating of land and water areas by the sun.

Aviation is so dependent upon that category of fluids called gases and the effect of forces and pressures acting upon gases, that a discussion of the subject of the atmosphere is important to a pilot. Data available about the atmosphere may determine whether a flight will succeed, or whether it will even become airborne. The various components of the air around the earth, the changes in temperatures and pressures at different levels above the earth, the properties of weather encountered by aircraft in flight, and many other detailed data points are considered in the preparation of flight plans.

Blaise Pascal (1623–1662) and Evangelista Torricelli (1608–1647) have been credited with developing the **barometer**, an instrument for measuring atmospheric pressure. The results of their experiments are still used today with very little improvement in design or knowledge. They determined that air has weight which changes as altitude is changed with respect to sea level. Today scientists are also interested in how the atmosphere affects the performance of the aircraft and its equipment.

2.1 Composition of the Atmosphere

The atmosphere is an envelope of air that surrounds the earth and rests upon its surface. It is as much a part of the earth as the seas or the land, but air differs from land and water as it is a mixture of gases. It has mass, weight, and indefinite shape.

The atmosphere consists of a mixture of various gases. (Figure 2.1) Pure, dry **air** is composed of approximately 78 percent nitrogen, 21 percent oxygen, and a 1 percent mixture of other gases, mostly argon. Some of these elements are heavier than others. The heavier elements, such as oxygen, settle to the surface of the earth, while the lighter elements are lifted up to the region of higher altitude. Most of the atmosphere's oxygen is contained below 35,000 feet altitude. Air, like a fluid, is able to flow and change shape when subjected to even minute changes in pressure because it lacks strong molecular cohesion. For example, gas completely fills any container into which it is placed, expanding or contracting to adjust its shape to the limits of the container. One of the most important factors of the atmosphere is **water vapor**, which varies in amounts from 0 to 5 percent by volume. It is present in three physical states: as a gas, liquid, and solid. The maximum amount of gaseous water vapor the air can hold is temperature dependent; the higher the temperature, the more water vapor it can hold. Water vapor remains invisibly suspended in varying amounts in the air until, through condensation, it grows to sufficient droplet or ice crystal size to form clouds or precipitation. Even when the atmosphere is apparently clear, it contains variable amounts of impurities, such as dust, smoke, volcanic ash, and salt particles.

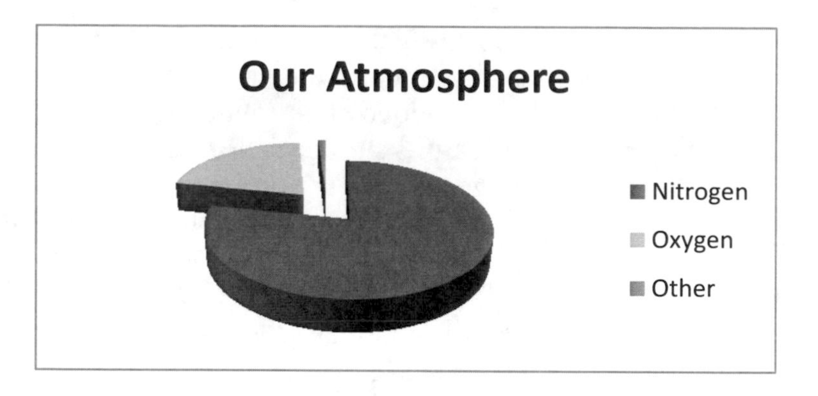

Figure 2.1. Our Atmosphere.

The atmosphere is a complex and ever changing mixture. Its ingredients vary from place to place and from day to day. In addition to a number of gases, it contains quantities of foreign matter such as pollen, dust, bacteria, soot, volcanic ash, spores, and dust from outer space. The composition of the air remains almost constant from sea level up to its highest level, but its density diminishes rapidly with altitude. Six miles up, for example, it is too thin to support respiration, and 12 miles up, there is not enough oxygen to support combustion, except in some specially designed turbine engine-powered airplanes.

At a point several hundred miles above the earth, some gas particles spray out into space, some are dragged by gravity and fall back into the ocean of air below, while others never return. Physicists disagree as to the boundaries of the outer fringes of the atmosphere. Some think it begins 240 miles above the earth and extends to 400 miles; others place its lower edge at 600 miles and its upper boundary at 6,000 miles. There are also certain nonconformities at various levels. Between 12 and 30 miles, high solar ultraviolet radiation reacts with oxygen molecules to produce a thin curtain of **ozone**, a very poisonous gas without which life on earth could not exist. This ozone filters out a portion of the sun's lethal ultraviolet rays, allowing only enough to come through to give us sunburn, kill bacteria, and prevent rickets.

At 50 to 65 miles up, most of the oxygen molecules begin to break down under solar radiation into free atoms, and to form hydroxy ions (OH) from water vapor. Also in this region, all the atoms become ionized.

Studies of the atmosphere have revealed that the temperature does not decrease uniformly with increasing altitude; instead it gets steadily colder up to a height of about 7 miles, where the rate of temperature change slows down abruptly and remains almost constant at −55° Centigrade (218° Kelvin) up to about 20 miles. Then the temperature begins to rise to a peak value of 77° Centigrade (350° Kelvin) at the 55 mile level. Thereafter it climbs steadily, reaching 2,270° Centigrade (2,543° Kelvin) at a height of 250 to 400 miles. From the 50 mile level upward, a human or any other living creature, without the protective cover of the **3-36** atmosphere, would be broiled on the side facing the sun and frozen on the other.

The atmosphere is divided into concentric layers or levels. Transition through these layers is gradual and without sharply defined boundaries. However, one boundary, the tropopause, exists between the first and second layer. The **tropopause** is defined as the point in the atmosphere at which the decrease in temperature (with increasing altitude) abruptly ceases. The four atmosphere layers are the troposphere, stratosphere, ionosphere, and the exosphere. The upper portion of the stratosphere is often called the chemosphere or ozonosphere, and the exosphere is also known as the mesosphere. The **troposphere** extends from the earth's surface to about 35,000 ft at middle latitudes, but varies from 28,000 ft at the poles to about 54,000 ft at the equator. (Figure 2.2)

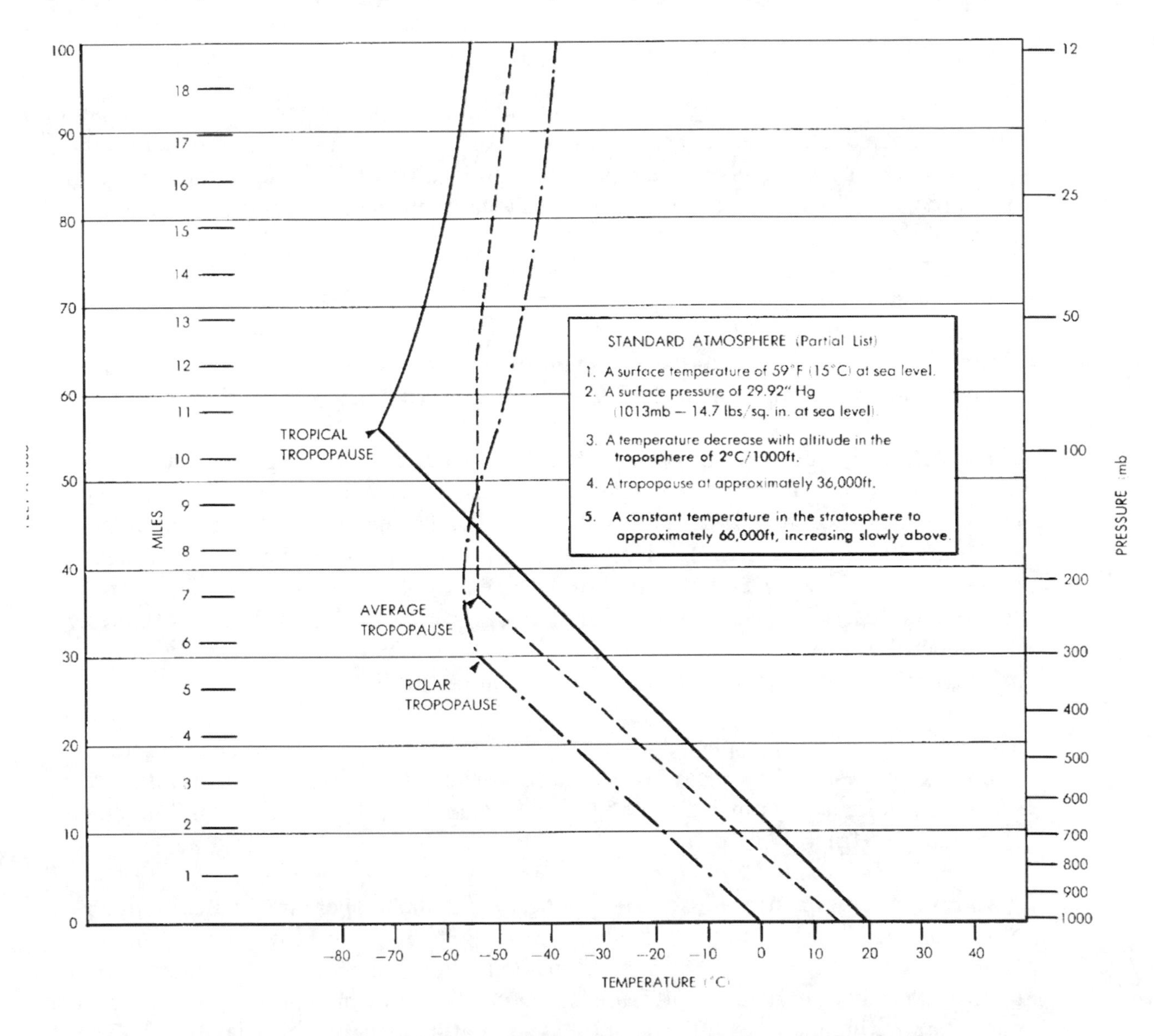

Figure 2.2. The Height of the Tropopause Varies with Latitude

The troposphere is characterized by large changes in temperature and humidity and by generally turbulent conditions. Nearly all cloud formations are within the troposphere. Approximately three-fourths of the total weight of the atmosphere is within the troposphere. The **stratosphere** extends from the upper limits of the troposphere (and the tropopause) to an average altitude of 60 miles. The **ionosphere** ranges from the 50 mile level to a level of 300 to 600 miles. Little is known about the characteristics of the ionosphere, but it is thought that many electrical phenomena occur there. Basically, this layer is characterized by the presence of ions and free electrons, and the ionization seems to increase with altitude and in successive layers. The **exosphere** (or mesosphere) is the outer layer of the atmosphere. It begins at an altitude of 600 miles and extends to the limits of the atmosphere. In this layer, the temperature is fairly constant at 2,500° Kelvin, and propagation of sound is thought to be impossible due to lack of molecular substance.

2.2 Atmospheric Pressure

Atmospheric pressure is one of the most important atmospheric parameters that aviators need to understand. Aviators need to know the differences among the air density, altimeter setting, sea level pressure, pressure altitude, density altitude, and constant pressure values. All of these pressures operationally impact aviators in significantly different ways.

Although there are various kinds of pressure, aviators are mainly concerned with atmospheric pressure. It is one of the basic factors in weather changes, helps to lift an aircraft, and actuates some of the important flight instruments. These instruments are the altimeter, airspeed indicator, vertical speed indicator, and manifold pressure gauge.

Air is very light, but it has mass and is affected by the attraction of gravity. Therefore, like any other substance, it has weight, and because of its weight, it has a force. Since it is a fluid substance, this force is exerted equally in all directions, and its effect on bodies within the air is called pressure. Under standard conditions at sea level, the average pressure exerted by the weight of the atmosphere is approximately 14.70 pounds per square inch (psi) of surface or 1,013.2 millibars (mb). Its thickness is limited; therefore, the higher the altitude, the less air there is above. For this reason, the weight of the atmosphere at 18,000 feet is one-half what it is at sea level.

The pressure of the atmosphere varies with time and location. Due to the changing atmospheric pressure, a standard reference was developed. The standard atmosphere at sea level is a surface temperature of 59°F or 15°C and a surface pressure of 29.92 inches of mercury ("Hg), or 1,013.2 mb.

A **standard temperature lapse rate** is one in which the temperature decreases at a rate of approximately 3.5°F or 2°C per thousand feet up to 36,000 feet which is approximately −65°F or −55°C. Above this point, the temperature is considered constant up to 80,000 feet. A standard pressure lapse rate is one in which pressure decreases at a rate of approximately 1 inch Hg per 1,000 feet of altitude gain up to 10,000 feet. The International Civil Aviation Organization (ICAO) has established this as a worldwide standard, and it is often referred to as **International Standard Atmosphere** (ISA) or

ICAO Standard Atmosphere. Any temperature or pressure that differs from the standard lapse rates is considered nonstandard temperature and pressure.

Atmospheric pressure is the force per unit area exerted by the weight of a column of air extending directly above a given fixed point. As a result of constant and complex air movements and changes in air mass characteristics, the weight of the air column is continually fluctuating. These changes in air weight, and therefore air pressure, are measured with pressure sensitive instruments call barometers.

The human body is under pressure, since it exists at the bottom of a sea of air. This pressure is due to the weight of the atmosphere. Figure 2.3 shows a standard day at sea level. If a 1-in^2 column of air extending to the top of the atmosphere was weighed, it would weigh 14.7 lb. That is why standard day atmospheric pressure is said to be 14.7 pounds per square inch (14.7 psi). Since atmospheric pressure at any altitude is due to the weight of air above it, pressure decreases with increased altitude. Obviously, the total weight of air above an area at 15,000 ft would be less than the total weight of the air above an area at 10,000 ft. Atmospheric pressure is often measured by a **mercury barometer**. A glass tube somewhat over 30 inches in length is sealed at one end and then filled with mercury. It is then inverted and the open end is placed in a dish of mercury. Immediately, the mercury level in the inverted tube will drop a short distance, leaving a small volume of mercury vapor at nearly zero absolute pressure in the tube just above the top of the liquid mercury column. Gravity acting on the mercury in the tube will try to make the mercury run out. Atmospheric pressure pushing down on the mercury in the open container tries to make the mercury stay in the tube. At some point these two forces (gravity and atmospheric pressure) will equilibrate and the mercury will stabilize at a certain height in the tube. Under standard day atmospheric conditions, the air in a 1-in^2 column extending to the top of the atmosphere would weigh 14.7 lb. A 1-in^2 column of mercury, 29.92 inches tall, would also weigh 14.7 lb. That is why 14.7 psi is equal to 29.92 "Hg. A second means of measuring atmospheric pressure is with an aneroid barometer. This mechanical instrument is a much better choice than a mercury barometer for use on airplanes. **Aneroid barometers (altimeters)** are used to indicate altitude in flight. The calibrations are made in thousands of feet above sea level, which is 29.92 "Hg, or 14.7 psi. At 10,000 feet above sea level, standard pressure is 20.58 "Hg, or 10.10 psi. Altimeters are calibrated so that if the pressure exerted by the atmosphere is 10.10 psi, the altimeter will point to 10,000 ft.

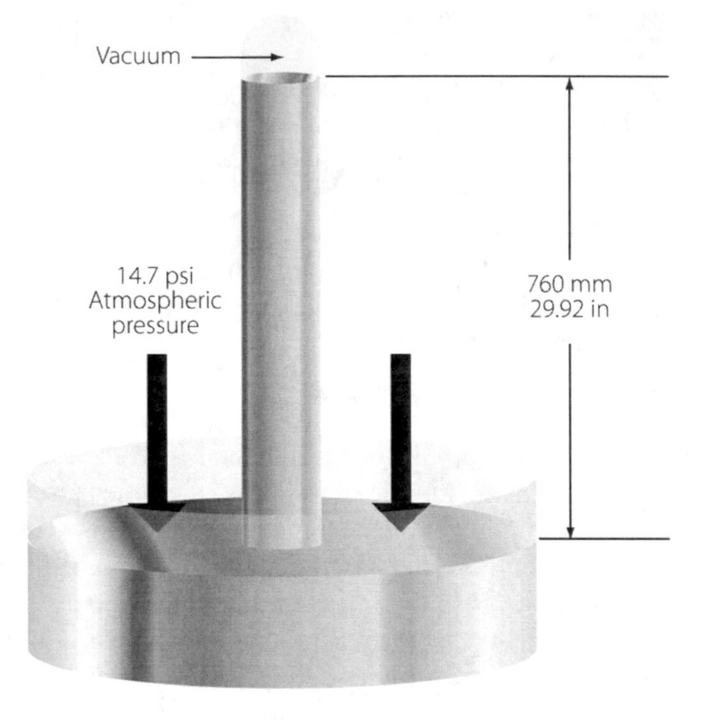

Figure 2.3. Our Atmosphere has Weight.

2.3 Atmospheric Density

Since both temperature and pressure decrease with altitude, it might appear that the density of the atmosphere would remain fairly constant with increased altitude. This is not true, however, because pressure drops more rapidly with increased altitude than does the temperature. The result is that density decreases with increased altitude. By use of the general gas law, studied earlier, it can be shown that for a particular gas, pressure and temperature determine the density. Since standard pressure and temperatures have been associated with each altitude, the density of the air at these standard temperatures and pressures must also be considered standard. Thus, a particular atmospheric density is associated with each altitude. This gives rise to the expression "**density altitude**," symbolized "Hd." A density altitude of 15,000 ft is the altitude at which the density is the same as that considered standard for 15,000 ft. Remember, however, that density altitude is not necessarily true altitude. For example, on a day when the atmospheric pressure is higher than standard and the temperature is lower than standard, the density which is standard at 10,000 ft might occur at 12,000 ft. In this case, at an actual altitude of 12,000 ft, we have air that has the same density as standard air at 10,000 ft. Density altitude is a calculated altitude obtained by correcting pressure altitude for temperature.

2.4 Pressure Altitude

Pressure altitude is the height above a **standard datum plane (SDP)**, which is a theoretical level where the weight of the atmosphere is 29.92 "Hg (1,013.2 mb) as measured by a barometer. An altimeter is essentially a sensitive barometer calibrated to indicate altitude in the standard atmosphere. If the altimeter is set for 29.92 "Hg SDP, the altitude indicated is the pressure altitude. As atmospheric pressure changes, the SDP may be below, at, or above sea level. Pressure altitude is important as a basis for determining airplane performance, as well as for assigning flight levels to airplanes operating at or above 18,000 feet.

The pressure altitude can be determined by either of two methods:

- Setting the barometric scale of the altimeter to 29.92 and reading the indicated altitude.
- Applying a correction factor to the indicated altitude according to the reported altimeter setting.

2.5 Density Altitude

SDP (standard datum plane) is a theoretical pressure altitude, but aircraft operate in a nonstandard atmosphere and the term density altitude is used for correlating aerodynamic performance in the nonstandard atmosphere. Density altitude is the vertical distance above sea level in the standard atmosphere at which a given density is to be found. The density of air has significant effects on the aircraft's performance because as air becomes less dense, it reduces:

- Power, because the engine takes in less air.
- Thrust, because a propeller is less efficient in thin air.
- Lift, because the thin air exerts less force on the airfoils.

Density altitude is pressure altitude corrected for nonstandard temperature. As the density of the air increases (lower density altitude), aircraft performance increases and conversely as air density decreases (higher density altitude), aircraft performance decreases. *A decrease in air density means a high density altitude; an increase in air density means a lower density altitude.* Density altitude is used in calculating aircraft performance, because under standard atmospheric conditions, air at each level in the atmosphere not only has a specific density, its pressure altitude and density altitude identify the same level.

The computation of density altitude involves consideration of pressure (pressure altitude) and temperature. Since aircraft performance data at any level are based upon air density under standard day conditions, such performance data apply to air density levels that may not be identical with altimeter indications. Under conditions higher or lower than standard, these levels cannot be determined directly from the altimeter.

Density altitude is determined by first finding pressure altitude, and then correcting this altitude for nonstandard temperature variations. Since density varies directly with pressure, and inversely with temperature, a given pressure altitude may exist for a wide range of temperature by allowing the density to vary. However, a known density occurs for any one temperature and pressure altitude. The density of the air has a pronounced effect on aircraft and engine performance.

Regardless of the actual altitude at which the aircraft is operating, it will perform as though it were operating at an altitude equal to the existing density altitude.

Air density is affected by changes in altitude, temperature, and humidity. **High density altitude** refers to thin air while **low density altitude** refers to dense air. The conditions that result in a high density altitude are high elevations, low atmospheric pressures, high temperatures, high humidity, or some combination of these factors. Lower elevations, high atmospheric pressure, low temperatures, and low humidity are more indicative of low density altitude.

2.6 Effect of Pressure on Density

Since air is a gas, it can be compressed or expanded. When air is compressed, a greater amount of air can occupy a given volume. Conversely, when pressure on a given volume of air is decreased, the air expands and occupies a greater space. At a lower pressure, the original column of air contains a smaller mass of air. The density is decreased because density is directly proportional to pressure. If the pressure is doubled, the density is doubled; if the pressure is lowered, the density is lowered. This statement is true only at a constant temperature.

2.7 Effect of Temperature on Density

Increasing the temperature of a substance decreases its density. Conversely, decreasing the temperature increases the density. Thus, the density of air varies inversely with temperature. This statement is true only at a constant pressure.

In the atmosphere, both temperature and pressure decrease with altitude, and have conflicting effects upon density. However, the fairly rapid drop in pressure as altitude is increased usually has the dominating effect. Hence, pilots can expect the density to decrease with altitude.

2.8 Effect of Humidity (Moisture) on Density

The preceding paragraphs refer to air that is perfectly dry. In reality, it is never completely dry. The small amount of water vapor suspended in the atmosphere may be almost negligible under certain conditions, but in other conditions humidity may become an important factor in the performance of an aircraft. Water vapor is lighter than air; consequently, moist air is lighter than dry air. Therefore, as the water content of the air increases, the air becomes less dense, increasing density altitude and decreasing performance. It is lightest or least dense when, in a given set of conditions, it contains the maximum amount of water vapor.

Humidity, also called relative humidity, refers to the amount of water vapor contained in the atmosphere, and is expressed as a percentage of the maximum amount of water vapor the air can hold. This amount varies with temperature. Warm air holds more water vapor, while colder air holds less. Perfectly dry air that contains no water vapor has a relative humidity of zero percent, while saturated air, which cannot hold any more water vapor, has a relative humidity of 100 percent. Humidity alone is usually not considered an important factor in calculating density altitude and aircraft performance, but it does contribute.

As temperature increases, the air can hold greater amounts of water vapor. When comparing two separate air masses, the first warm and moist (both qualities tending to lighten the air) and the second cold and dry (both qualities making it heavier), the first must be less dense than the second. Pressure, temperature, and humidity have a great influence on aircraft performance because of their effect upon density. There are no rules of thumb that can be easily conveyed but the effect of humidity can be determined using online formulas. In the first case, the pressure is needed at the altitude for which density altitude is being sought. Using Figure 2.2, select the barometric pressure closest to the associated altitude. As an example, the pressure at 8,000 feet is 22.22 "Hg. Using the National Oceanic and Atmospheric Administration (NOAA) website (http://www.srh.noaa.gov/elp/wxcalc/densityaltitude.html) for density altitude, enter the 22.22 for 8,000 feet in the station pressure window. Enter a temperature of 80° and a dew point of 75°. The result is a density altitude of 11,564 feet. With no humidity, the density altitude would be almost 500 feet lower.

Another site (http://wahiduddin.net/calc/density_altitude.htm) provides a more straightforward method of determining the effects of humidity on density altitude without using additional interpretive charts. In any case, the effects of humidity on density altitude include a decrease in overall performance in high humidity conditions.

2.9 Water Content of the Atmosphere

In the troposphere, the air is rarely completely dry. It contains water vapor in one of two forms: (1) fog or (2) water vapor. **Fog** consists of minute droplets of water held in suspension by the air. **Clouds** are composed of fog. The height to which some clouds extend is a good indication of the presence of water in the atmosphere almost up to the stratosphere.

As a result of evaporation, the atmosphere always contains some moisture in the form of water vapor. The moisture in the air is called the humidity of the air. Moisture does not consist of tiny particles of liquid held in suspension in the air as in the case of fog, but is an invisible vapor truly as gaseous as the air with which it mixes.

Fog and humidity both affect the performance of an aircraft. In flight, at cruising power, the effects are small and receive no consideration. During takeoff, however, humidity has important effects. Two things are done to compensate for the effects of humidity on takeoff performance. Since humid air is less dense than dry air, the allowable takeoff gross weight of an aircraft is generally reduced for operation in areas that are consistently humid. Second, because the power output of reciprocating engines is decreased by humidity, the manifold pressure may need to be increased above that recommended for takeoff in dry air in order to obtain the same power output.

Engine power output is calculated on dry air. Since water vapor is incombustible, its pressure in the atmosphere is a total loss as far as contributing to power output. The mixture of water vapor and air is drawn through the carburetor, and fuel is metered into it as though it were all air. This mixture of water vapor, air, and fuel enters the combustion chamber where it is ignited. Since the water vapor will not burn, the effective fuel/air ratio is enriched and the engine operates as though it were on an excessively rich mixture. The resulting horsepower loss under humid conditions can therefore be attributed to the loss in volumetric efficiency due to displaced air, and the incomplete combustion due to an excessively rich fuel/air mixture. The reduction in power that can be expected from humidity is usually given in charts in the flight manual. There are several types of charts in use. Some merely show the expected reduction in power due to humidity; others show the boost in manifold pressure necessary to restore full takeoff power. The effect of fog on the performance of an engine is very noticeable, particularly on engines with high compression ratios. Normally, some detonation will occur during acceleration, due to the high BMEP (brake mean effective pressures) developed. However, on a foggy day it is difficult to cause detonation to occur. The explanation of this lies in the fact that fog consists of particles of water that have not vaporized. When these particles enter the cylinders, they absorb a tremendous amount of heat energy in the process of vaporizing. The temperature is thus lowered, and the decrease is sufficient to prevent detonation. Fog will generally cause a decrease in horsepower output. However, with a supercharged engine, it will be possible to use higher manifold pressures without danger of detonation.

2.9.1 Absolute Humidity

Absolute humidity is the actual amount of the water vapor in a mixture of air and water. It is expressed either in grams per cubic meter or pounds per cubic foot. The amount of water vapor that can be present in the air is dependent upon the temperature and pressure. The higher the temperatures, the more water vapor the air is capable of holding, assuming constant pressure. When air has all the water vapor it can hold at the prevailing temperature and pressure, it is said to be **saturated**.

2.9.2 Relative Humidity

Relative humidity is the ratio of the amount of water vapor actually present in the atmosphere to the amount that would be present if the air were saturated at the prevailing temperature and pressure. This ratio is usually multiplied by 100 and expressed as a percentage. Suppose, for example, that a weather report includes the information that the temperature is 75°F and the relative humidity is 56 percent. This indicates that the air holds 56 percent of the water vapor required to saturate it at 75°F. If the temperature drops and the absolute humidity remains constant, the relative humidity will increase. This is because less water vapor is required to saturate the air at the lower temperature.

2.9.3 Dew Point

The **dew point** is the temperature to which humid air must be cooled at constant pressure to become saturated. If the temperature drops below the dew point, condensation occurs. People who wear eyeglasses have experience going from cold outside air into a warm room and having moisture collect quickly on their glasses. This happens because the glasses were below the dew point temperature of the air in the room. The air immediately in contact with the glasses was cooled below its dew point temperature, and some of the water vapor was condensed out. This principle is applied in determining the dew point. A vessel is cooled until water vapor begins to condense on its surface. The temperature at which this occurs is the dew point.

2.9.4 Vapor Pressure

Vapor pressure is the portion of atmospheric pressure that is exerted by the moisture in the air (expressed in tenths of an inch of mercury). The dew point for a given condition depends on the amount of water pressure present; thus, a direct relationship exists between the vapor pressure and the dew point.

2.10 Standard Atmosphere

If the performance of an aircraft is computed, either through flight tests or wind tunnel tests, some standard reference condition must be determined first in order to compare results with those of similar tests. The conditions in the atmosphere vary continuously, and it is generally not possible to obtain exactly the same set of conditions on two different days or even on two successive flights. For this reason, a set group of standards must be used as a point of reference. The set of standard conditions presently used in the United States is known as the **U.S. Standard Atmosphere**. The standard atmosphere approximates the average conditions existing at 40° latitude, and is determined on the basis of the following assumptions. The standard sea level conditions are:

Pressure at 0 altitude (P0) = 29.92 "Hg
Temperature at 0 altitude (T0) = 15°C or 59°F
Gravity at 0 altitude (G0) = 32.174 fps/s

The U.S. Standard Atmosphere is in agreement with the International Civil Aviation Organization (ICAO) Standard Atmosphere over their common altitude range. The ICAO Standard Atmosphere has been adopted as standard by most of the principal nations of the world.

Note: Figures in Chapter 2 were taken from Aerodynamics for Pilot's, ATC Pamphlet 51-3

Chapter 3

The Four Forces Acting on an Airplane

A **force** may be thought of as a push or pull in a specific direction. A force is a vector quantity so a force has both a magnitude and a direction. When describing forces, we have to specify both the magnitude and the direction of the force. Figure 3.1 shows the forces that act on an airplane in flight: weight, lift, drag, and thrust.

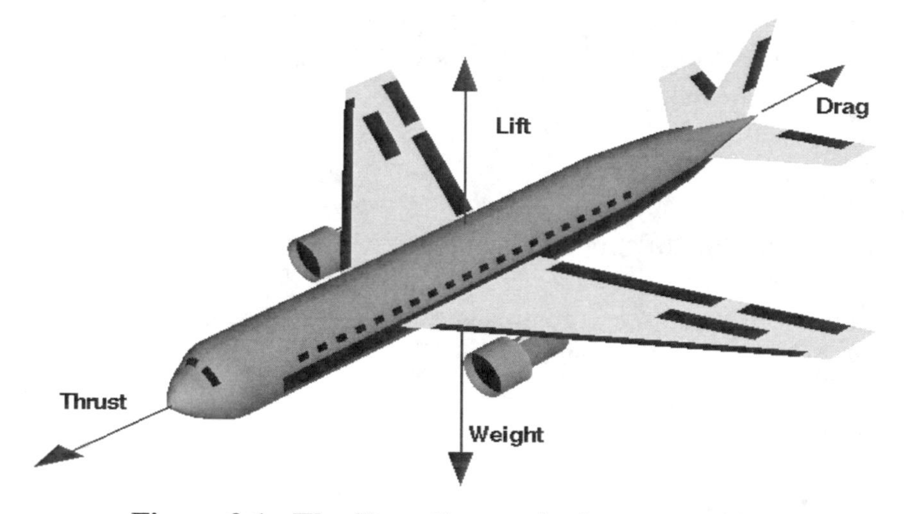

Figure 3.1. The Four Forces Acting on an Airplane.
Source: NASA Beginner's Guide to Aerodynamics (BGA)

3.1 Weight

Weight is a force that is always directed toward the center of the earth. The magnitude of the weight depends on the mass of all the airplane parts, plus the amount of fuel, plus any payload on board (people, baggage, freight, etc.). The weight is distributed throughout the airplane. But we can often think of it as collected and acting through a single point called the center of gravity. In flight, the airplane rotates about the **center of gravity (CG)**.

Flying encompasses two major problems; overcoming the weight of an object by some opposing force, and controlling the object in flight. Both of these problems are related to the object's weight and the location of the center of gravity. During a flight, an airplane's weight constantly changes as the aircraft consumes fuel. The distribution of the weight and the center of gravity also changes. So the pilot must constantly adjust the controls to keep the airplane balanced, or **trimmed**.

3.2 Lift

To overcome the weight force, airplanes generate an opposing force called **lift**. Lift is generated by the motion of the airplane through the air and is an aerodynamic force. "*Aero*" stands for the air, and "*dynamic*" denotes motion. Lift is directed *perpendicular* to the flight direction. The magnitude of the lift depends on several factors including the shape, size, and velocity of the aircraft. As with weight, each part of the aircraft contributes to the aircraft lift force. Most of the lift is generated by the wings. Aircraft lift acts through a single point called the center of pressure. The **center of pressure** is defined just like the center of gravity, but using the *pressure* distribution around the body instead of the *weight* distribution.

Lift opposes the downward force of weight, is produced by the dynamic effect of the air acting on the airfoil, and acts perpendicular to the flightpath through the center of lift. The distribution of lift around the aircraft is important for solving the control problem. Aerodynamic surfaces are used to control the aircraft in roll, pitch, and yaw. Lift is explained in greater detail in Chapter 4.

3.3 Drag

As the airplane moves through the air, there is another aerodynamic force present. The air resists the motion of the aircraft and the resistance force is called drag. Drag is directed *along and opposed* to the flight direction. Like lift, there are many factors that affect the magnitude of the drag force, including the shape of the aircraft, the "stickiness" of the air, and the velocity of the aircraft. Like lift, we collect all of the individual components' drags and combine them into a single aircraft drag magnitude. And like lift, drag acts through the aircraft center of pressure.

Drag is the force that resists movement of an aircraft through the air. There are two basic types: **parasite drag** and **induced drag**. The first is called parasite because it in no way functions to aid flight, while the second, induced drag, is a result of an airfoil developing lift. Drag is discussed in more detail in Chapter 5.

3.4 Thrust

To overcome drag, airplanes use a propulsion system to generate a force called thrust. The direction of the thrust force depends on how the engines are attached to the aircraft. In the figure shown above, two turbine engines are located under the wings, parallel to the body, with thrust acting along the body centerline. On some aircraft, such as the Harrier, the thrust direction can be varied to help the airplane take off in a very short distance. The magnitude of the thrust depends on many factors associated with the propulsion system, including the type of engine, the number of engines, and the throttle setting.

Thrust is the forward force produced by the power plant/ propeller or rotor. It opposes or overcomes the force of drag. As a general rule, it acts parallel to the longitudinal axis. However, this is not always the case, as explained later.

For gas turbine engines, aircraft thrust is a reaction to the hot gas rushing out of the nozzle. The hot gas typically goes out the back, but the thrust pushes toward the front. This action/reaction is explained by Newton's third law of motion.

The motion of the airplane through the air depends on the relative strengths and directions of the forces shown above. If the forces are balanced, the aircraft cruises at constant velocity. If the forces are unbalanced, the aircraft accelerates in the direction of the largest force.

For an aircraft to move, thrust must be exerted and be greater than drag. The aircraft will continue to move and gain speed until thrust and drag are equal. In order to maintain a constant airspeed, thrust and drag must remain equal, just as lift and weight must be equal to maintain a constant altitude. If, in level flight, the engine power is reduced, the thrust is lessened, and the aircraft slows down. As long as the thrust is less than the drag, the aircraft continues to decelerate until its airspeed is insufficient to support it in the air.

Likewise, if the engine power is increased, thrust becomes greater than drag and the airspeed increases. As long as the thrust continues to be greater than the drag, the aircraft continues to accelerate. When drag equals thrust, the aircraft flies at a constant airspeed.

3.5 Steady Flight

In steady flight, the sum of these opposing forces is always zero. There can be no unbalanced forces in steady, straight flight based upon Newton's Third Law, which states that for every action or force there is an equal, but opposite, reaction or force. This is true whether flying level or when climbing or descending.

It does not mean the four forces are equal. It means the opposing forces are equal to, and thereby cancel, the effects of each other. In Figure 3.1 the force vectors of thrust, drag, lift, and weight appear to be equal in value. The usual explanation states (without stipulating that thrust and drag do not equal weight and lift) that thrust equals drag and lift equals weight. Although basically true, this statement can be misleading. It should be understood that in straight, level, unaccelerated flight, it is true that the opposing lift/weight forces are equal. They are also greater than the opposing forces of thrust/drag that are equal only to each other. Therefore, in steady flight, the sum of all upward forces (not just lift) equals the sum of all downward forces (not just weight). Additionally, the sum of all forward forces (not just thrust) equals the sum of all backward forces (not just drag). (Figure 3.2).

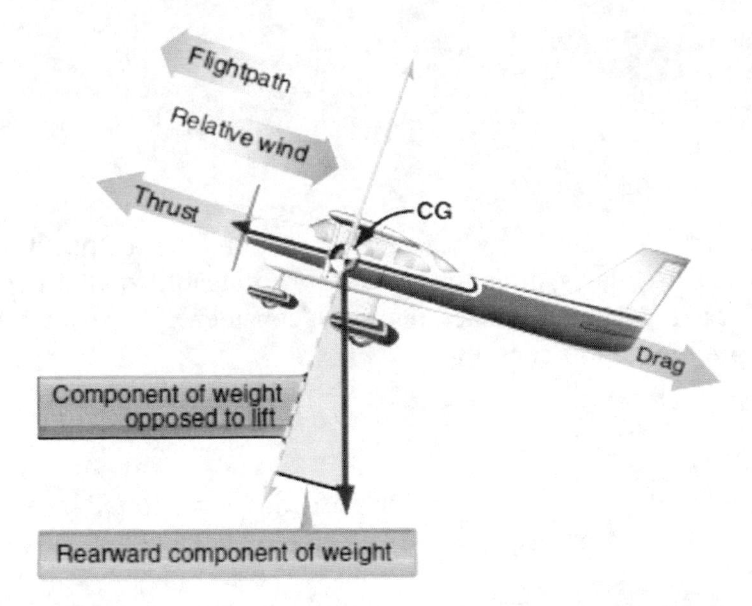

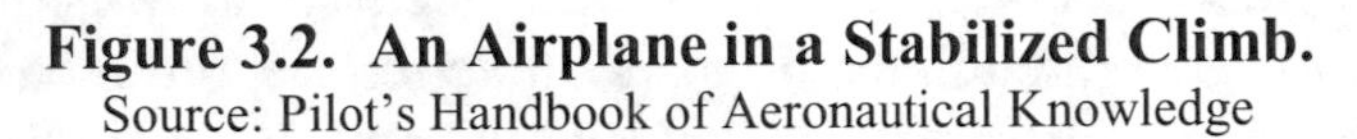

Figure 3.2. An Airplane in a Stabilized Climb.
Source: Pilot's Handbook of Aeronautical Knowledge

This refinement of the old "thrust equals drag; lift equals weight" formula explains that a portion of thrust is directed upward in climbs and acts as if it were lift, while a portion of weight is directed backward and acts as if it were drag.

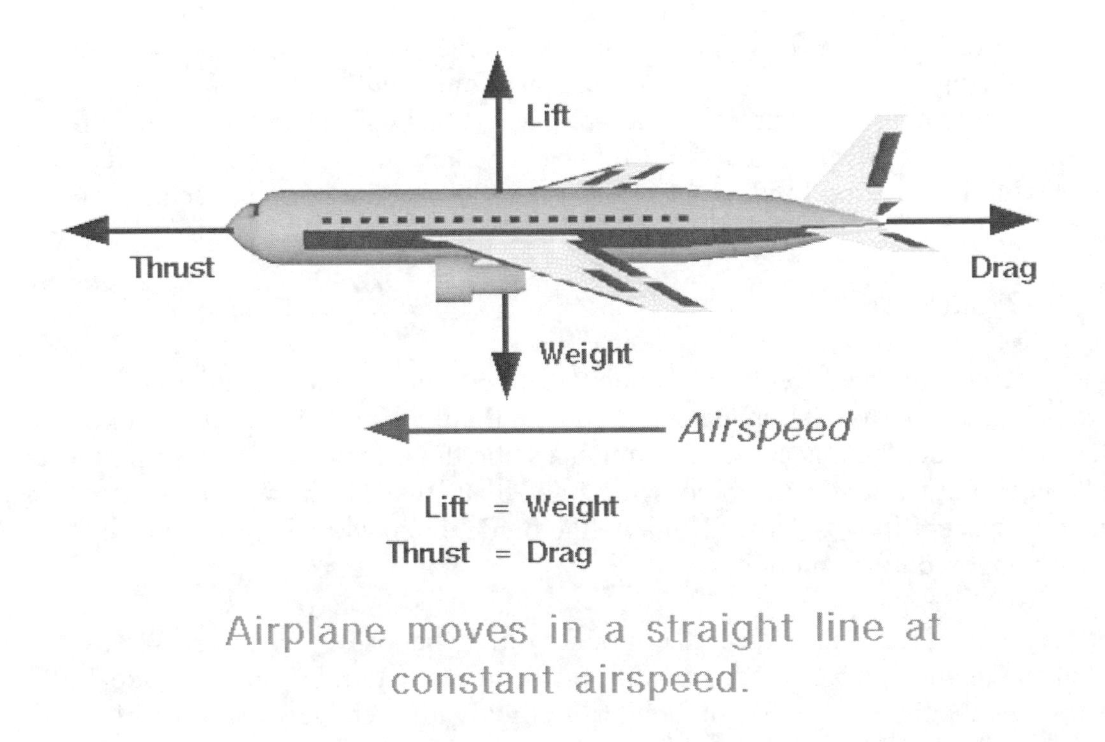

Figure 3.3. An Airplane in Stabilized Cruise.
Source: NASA Beginner's Guide to Aerodynamics (BGA)

Chapter 4

Lift

The formulation of lift has historically been the adaptation over the past few centuries of basic physical laws. These laws, although seemingly applicable to all aspects of lift, do not answer how lift is formulated. In fact, one must consider the many airfoils that are symmetrical, yet produce significant lift.

4.1 How is lift generated?

Lift is the force that holds an aircraft in the air. Lift can be generated by any part of the airplane, but most of the lift on an airplane is generated by the wings (airfoil).

Lift occurs when a flow of gas is turned by a solid object. The flow is turned in one direction, and the lift is generated in the opposite direction, according to Newton's third law of action and reaction. Because air is a gas and the molecules are free to move about, any solid surface can deflect a flow. Additionally there is a small, but significant, difference in static pressure between the upper and lower surfaces of an airfoil. For an airfoil, both the upper and lower surfaces contribute to the generation of lift.

4.1.1 No Fluid, No Lift

Lift is a mechanical force. It is generated by the interaction and contact of a solid body with a fluid (liquid or gas). It is not generated by a force field, in the sense of a gravitational field, or an electromagnetic field, where one object can affect another object without being in physical contact. For lift to be generated, the solid body must be in contact with the fluid: no fluid, no lift. Without air, there is no lift generated by the wings.

4.1.2 No Motion, No Lift

Lift is generated by the difference in velocity between the solid object and the fluid. There must be motion between the object and the fluid: no motion, no lift. It makes no difference whether the object moves through a static fluid, or the fluid moves past a static solid object. For a kite, the lift force is generated by the wind blowing over a surface that is fixed in space. Lift acts perpendicular to the fluid motion while drag acts in the direction opposed to the motion.

There are many factors that affect the amount of lift generated by an object. We can express the relation between these factors and the generated lift by a mathematical lift equation. At the time of the Wright brothers a slightly different version of the lift equation was used, but both old and new versions of the equation express the dependence of lift on the square of the velocity, the surface area, the shape of the object, inclination of the object to the flow, and the properties of the air.

4.2 Newton's Laws and Lift

The fundamental physical laws governing the forces acting upon an aircraft in flight were adopted from postulated theories developed before any human successfully flew an aircraft. The use of these physical laws grew out of the Scientific Revolution, which began in Europe in the 1600s. Driven by the belief that the universe operated in a predictable manner open to human understanding, many philosophers, mathematicians, natural scientists, and inventors spent their lives unlocking the secrets of the universe. One of the best known was Sir Isaac Newton, who not only formulated the law of universal gravitation, but also described the three basic laws of motion.

Newton's First Law: "Every object persists in its state of rest or uniform motion in a straight line unless it is compelled to change that state by forces impressed on it." This means that nothing starts or stops moving until some outside force causes it to do so. An aircraft at rest on the ramp remains at rest unless a force strong enough to overcome its inertia is applied. Once it is moving, its inertia keeps it moving, subject to the various other forces acting on it. These forces may add to its motion, slow it down, or change its direction.

Newton's Second Law: "Force is equal to the change in momentum per change in time. For a constant mass, force equals mass times acceleration." When a body is acted upon by a constant force, its resulting acceleration is inversely proportional to the mass of the body and is directly proportional to the applied force. This takes into account the factors involved in overcoming Newton's first law. It covers both changes in direction and speed, including starting up from rest (positive acceleration) and coming to a stop (negative acceleration or deceleration).

Newton's Third Law: "For every action, there is an equal and opposite reaction." In an airplane, the propeller moves and pushes back the air; consequently, the air pushes the propeller (and thus the airplane) in the opposite direction—forward. In a jet airplane, the engine pushes a blast of hot gases backward; the force of equal and opposite reaction pushes against the engine and forces the airplane forward. Additionally, the wings of the airplane turn the air. The flow is turned in one direction, and the lift is generated in the opposite direction, according to Newton's third law of action and reaction. For an airfoil, both the upper and lower surfaces contribute to the flow turning.

4.3 Bernoulli's Law and Lift

In the 1700s, a half-century after Newton formulated his laws, Daniel Bernoulli (1700–1782), a Swiss mathematician, explained how the pressure of a moving fluid (liquid or gas) varies with its speed of motion. **Bernoulli's principle** states that as the velocity of a moving fluid (liquid or gas) increases, the pressure within the fluid decreases. The equation states that the static pressure (P_S) in the flow plus the dynamic pressure (P_D) in the flow equals the total pressure (P_T).

A distinguishing feature of subsonic airflow is that changes in pressure and velocity take place with small and negligible changes in density. For this reason the study of subsonic airflow can be simplified by neglecting the variation of density in the flow and assuming the flow to be incompressible. When we study transonic, supersonic, and hypersonic flight, we do not assume the flow to be incompressible. If the flow through the tube of Figure 4.1 is considered to be fully subsonic, the density of the airstream is essentially constant at all stations along the length of the tube.

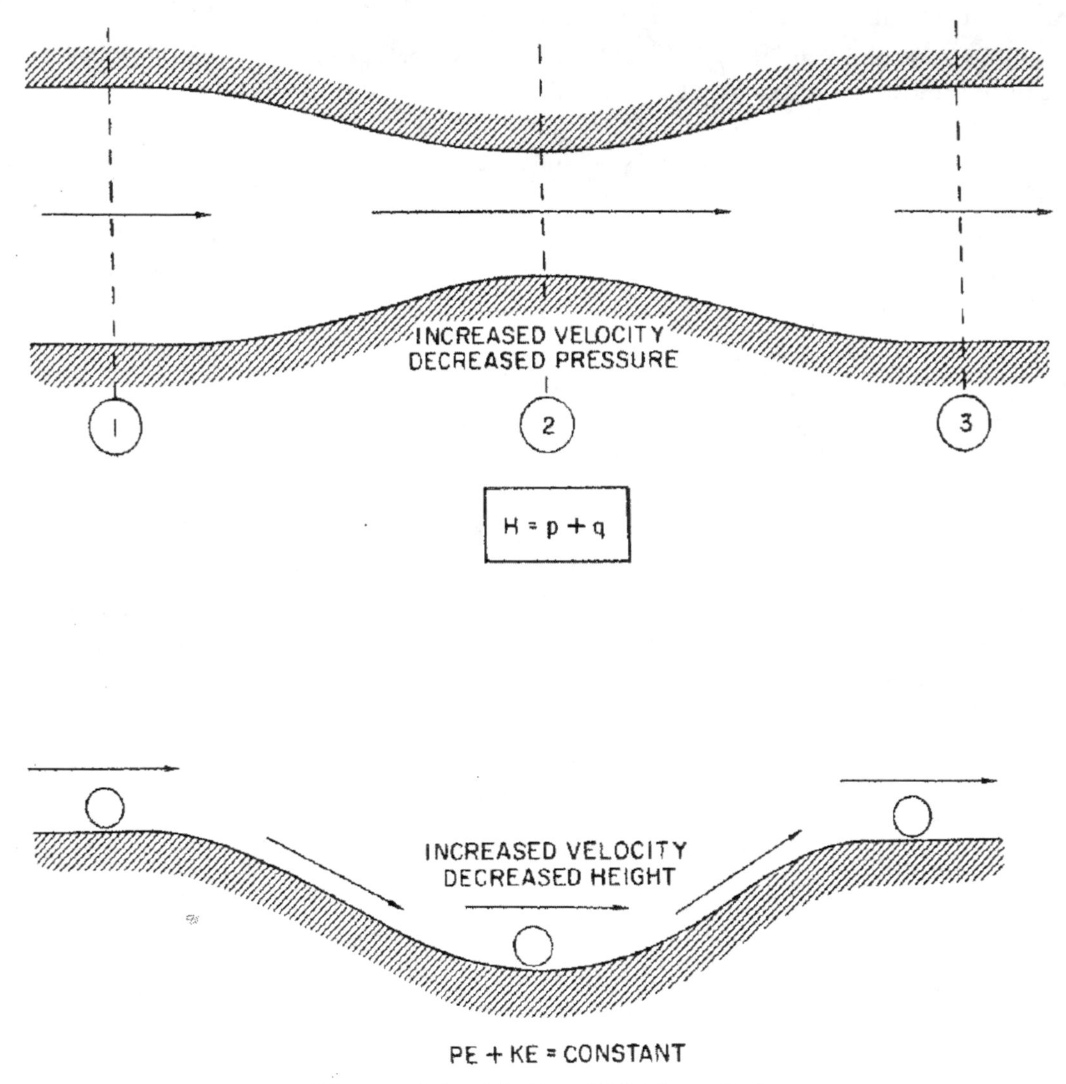

Figure 4.1. Airflow Within a Tube
Source: Naval Aviator's Guide to Aerodynamics

If the density of the flow remains constant, static pressure and velocity are the variable quantities. As the flow approaches the constriction of station 2, the velocity must increase to maintain the same mass flow. As the velocity increases, the static pressure will decrease and the decrease in static pressure accompanies the increase in velocity.

The total energy of the airstream in the tube is unchanged. However, the airstream energy may be in two forms. The airstream may have a potential energy (PE in the figure) which is related by the static pressure and a kinetic energy (KE in the figure) by virtue of mass and motion. As the total energy is unchanged, an increase in velocity (kinetic energy) will be accompanied by a decrease in static pressure (potential energy).

4.3.1 Bernoulli's Principle of Differential Pressure

A practical application of Bernoulli's principle is the Venturi tube. The **Venturi tube** has an air inlet that narrows to a throat (constricted point) and an outlet section that increases in diameter toward the rear. The diameter of the outlet is the same as that of the inlet. At the throat, the airflow speeds up and the pressure decreases; at the outlet, the airflow slows and the pressure increases.

Since air is recognized as a body and it is accepted that it must follow the above laws, one can begin to see how and why an airplane wing develops lift. As the wing moves through the air, the flow of air across the curved top surface increases in velocity, creating a low-pressure area.

4.4 Magnus Effect

In 1852, the German physicist and chemist Heinrich Gustav Magnus (1802–1870) made experimental studies of the aerodynamic forces on spinning spheres and cylinders. (The effect had already been mentioned by Newton in 1672, apparently in regard to spheres or tennis balls). These experiments led to the discovery of the Magnus effect, which helps explain the theory of lift.

4.4.1. Flow of Air Against a Nonrotating Cylinder

If air flows against a cylinder that is not rotating, the flow of air above and below the cylinder is identical and the forces are the same. (Figure 4.2) These typical airflow patterns exemplify the relationship of static pressure and velocity defined by Bernoulli. Any object placed in an airstream will have the air to impact or stagnate at some point near the leading edge. The pressure at this point of stagnation will be an absolute static pressure equal to the total pressure of the airstream. In other words, the static pressure at the stagnation point will be greater than the atmospheric pressure by the amount of the dynamic pressure of the airstream. As the flow divides and proceeds around the object, the increases in local velocity produce decreases in static pressure. When the streamlines contract and are close together, the high local velocities exist. When the streamlines expand and are far apart, low local velocities exist. At the forward stagnation point the local velocity is zero and the maximum positive pressure results. As the flow proceeds from the forward stagnation point, the velocity increases as shown by the change in streamlines. The local velocities reach a maximum at the upper and lower extremities and a peak suction pressure is produced at these points on the cylinder. (Note: **Positive pressures** are pressures above atmospheric and **negative or suction pressures** are less than atmospheric.) As the flow continues aft from the peak suction pressure, the diverging streamlines indicate decreasing local velocities and increasing local pressures.

If friction and compressibility effects are not considered, the velocity would decrease to zero at the aft stagnation point and the full stagnation pressure would be recovered. The pressure distribution for the cylinder in perfect fluid distribution for the cylinder in a perfect fluid flow would be symmetrical and no net force (lift or drag) would result. Of course, the relationship between static pressure and velocity along the surface is defined by Bernoulli's equation.

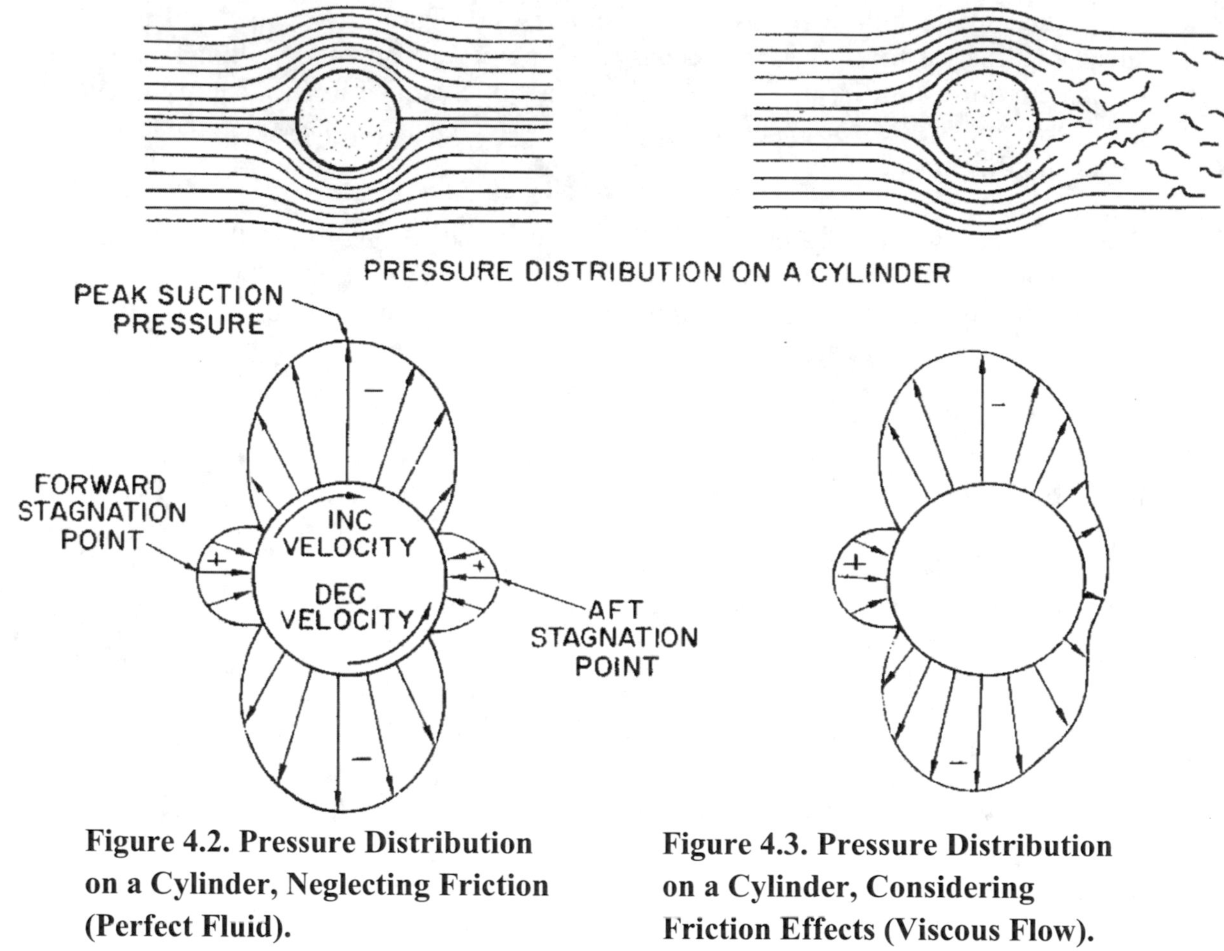

Figure 4.2. Pressure Distribution on a Cylinder, Neglecting Friction (Perfect Fluid).

Figure 4.3. Pressure Distribution on a Cylinder, Considering Friction Effects (Viscous Flow).

Source: Aerodynamics for Naval Aviators

The flow pattern for the cylinder in an actual fluid demonstrates the effect of friction or viscosity. (Figure 4.3) The viscosity of air produces a thin layer of retarded flow immediately adjacent to the surface. The energy expended in this "boundary layer" can alter the pressure distribution and destroy the symmetry of the pattern. The force imbalance caused by the change in pressure distribution creates a drag force which is in addition to the drag due to skin friction.

4.4.2 A Rotating Cylinder in a Motionless Fluid

In Figure 4.4, the cylinder is rotated clockwise and observed from the side while immersed in a fluid. The rotation of the cylinder affects the fluid surrounding the cylinder. The flow around the rotating cylinder differs from the flow around a stationary cylinder due to resistance caused by two factors: viscosity and friction.

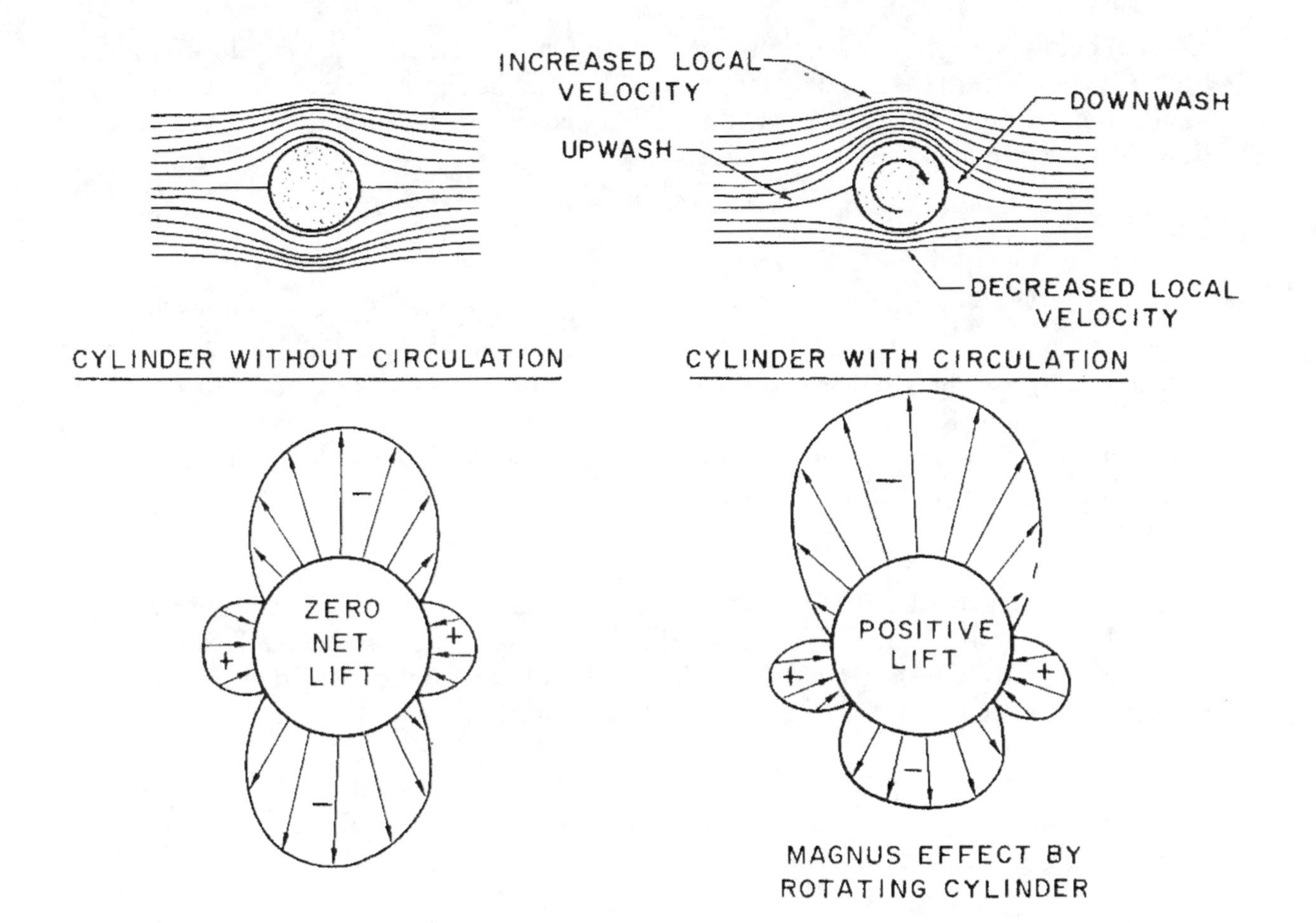

Figure 4.4. The Magnus Effect.
Source: Aerodynamics for Naval Aviators

If the cylinder is given a clockwise rotation and induces a rotational or circulatory flow, a distinct change takes place in the streamline pattern and pressure distribution. The velocities due to the vortex of circulatory flow cause increased local velocity on the upper surface of the cylinder and decreased local velocity on the lower surface of the cylinder. Also, the circulatory flow produces an upwash immediately ahead and downwash immediately behind the cylinder and both force and aft stagnation points are lowered.

The effect of the addition of circulatory flow is appreciated by the change in the pressure distribution on the cylinder. The increased local velocity on the upper surface causes an increase in upper surface suction while the decreased local velocity on the lower surface causes a decrease in lower surface suction. As a result, the cylinder with circulation will produce a net lift. This mechanically induced circulation is called the **Magnus effect**. The Magnus effect illustrates the relationship between circulation and lift and is important to golfers, baseball and tennis players, as well as pilots and aeronautical engineers. The curvature of the flightpath of a golf ball or baseball requires an unbalance of force which is created by rotation of the ball.

When the free stream flow is added to this circular flow, the resulting flow has a net turning and produces a force. On Figure 4.4, the cylinder spins clockwise, so the free stream below the cylinder is opposed by the circular flow; the free stream flow over the top of the cylinder is assisted by the circular flow. In the figure we can see that the streamlines around the cylinder are distorted because of the spinning. The net turning of the flow has produced an upward force.

4.5 Viscosity

Viscosity is the property of a fluid or semi fluid that causes it to resist flowing. This resistance to flow is measurable due to the molecular tendency of fluids to adhere to each other to some extent. High-viscosity fluids resist flow; low-viscosity fluids flow easily.

Similar amounts of oil and water poured down two identical ramps demonstrate the difference in viscosity. The water seems to flow freely while the oil flows much more slowly.

Since molecular resistance to motion underlies viscosity, grease is very viscous because its molecules resist flow. Hot lava is another example of a viscous fluid. All fluids are viscous and have a resistance to flow whether this resistance is observed or not. Air is an example of a fluid whose viscosity cannot be observed.

Since air has viscosity properties, it will resist flow to some extent. In the case of the rotating cylinder within an immersed fluid (oil, water, or air), the fluid (no matter what it is) resists flowing over the cylinder's surface.

4.6 Friction

Friction is the second factor at work when a fluid flows around a rotating cylinder. **Friction** is the resistance one surface or object encounters when moving over another and exists between a fluid and the surface over which it flows.

If identical fluids are poured down the ramp, they flow in the same manner and at the same speed. If one ramp's surface is coated with small pebbles, the flow down the two ramps differs significantly. The rough surface ramp impedes the flow of the fluid due to resistance from the surface (friction). It is important to remember that all surfaces, no matter how smooth they appear, are not completely smooth and impede the flow of a fluid. Both the surface of a wing and the rotating cylinder have a certain roughness, albeit at a microscopic level, causing resistance for a fluid to flow. This reduction in velocity of the airflow about a surface is caused by skin friction or drag.

When passing over a surface, molecules actually adhere (stick or cling) to the surface, illustrated by the rotating cylinder in a fluid that is not moving. Thus, in the case of the rotating cylinder, air particles near the surface that resist motion have a relative velocity near zero. The roughness of the surface impedes their motion. Additionally, due to the viscosity of the fluid, the molecules on the surface **entrain**, or pull, the surrounding flow above it in the direction of rotation due to the adhesion of the fluid to itself.

There is also a difference in flow around the rotating cylinder and in flow around a nonrotating cylinder. The molecules at the surface of the rotating cylinder are not in motion relative to the cylinder; they are moving clockwise with the cylinder. Due to viscosity, these molecules entrain others above them resulting in an increase in fluid flow in the clockwise direction. Substituting air for other fluids results in a higher velocity of air movement above the cylinder simply because more molecules are moving in a clockwise direction.

4.7 Aerodynamic Forces on an Airfoil

An **airfoil** is a structure designed to obtain reaction upon its surface from the air through which it moves or that moves past such a structure. Air acts in various ways when submitted to different pressures and velocities; but this discussion is confined to the parts of an aircraft that a pilot is most concerned with in flight—namely, the airfoils designed to produce lift. By looking at a typical airfoil profile, such as the cross section of a wing, one can see several obvious characteristics of design. (Figure 4.5) On most subsonic airfoil there is a difference in the curvatures (called **cambers**) of the upper and lower surfaces of the airfoil. The camber of the upper surface is more pronounced than that of the lower surface, which is usually somewhat flat.

An airfoil is constructed in such a way that its shape takes advantage of the air's response to certain physical laws. This develops two actions from the air mass: a positive pressure lifting action from the air mass below the wing, and a negative pressure lifting action from lowered pressure above the wing.

As the airstream strikes the relatively flat lower surface of a wing or rotor blade when inclined at a small angle to its direction of motion, the air is forced to rebound downward, causing an upward reaction in positive lift. At the same time, the airstream striking the upper curved section of the leading edge is deflected upward. An airfoil is shaped to cause an action on the air, and forces air downward, which provides an equal reaction from the air, forcing the airfoil upward. If a wing is constructed in such form that it causes a lift force greater than the weight of the aircraft, the aircraft will fly.

If all the lift required were obtained merely from the deflection of air by the lower surface of the wing, an aircraft would only need a flat wing like a kite. However, the balance of the lift needed to support the aircraft comes from the flow of air above the wing. Herein lies the key to flight.

It is neither accurate nor useful to assign specific values to the percentage of lift generated by the upper surface of an airfoil versus that generated by the lower surface. These are not constant values and vary, not only with flight conditions, but also with different wing designs.

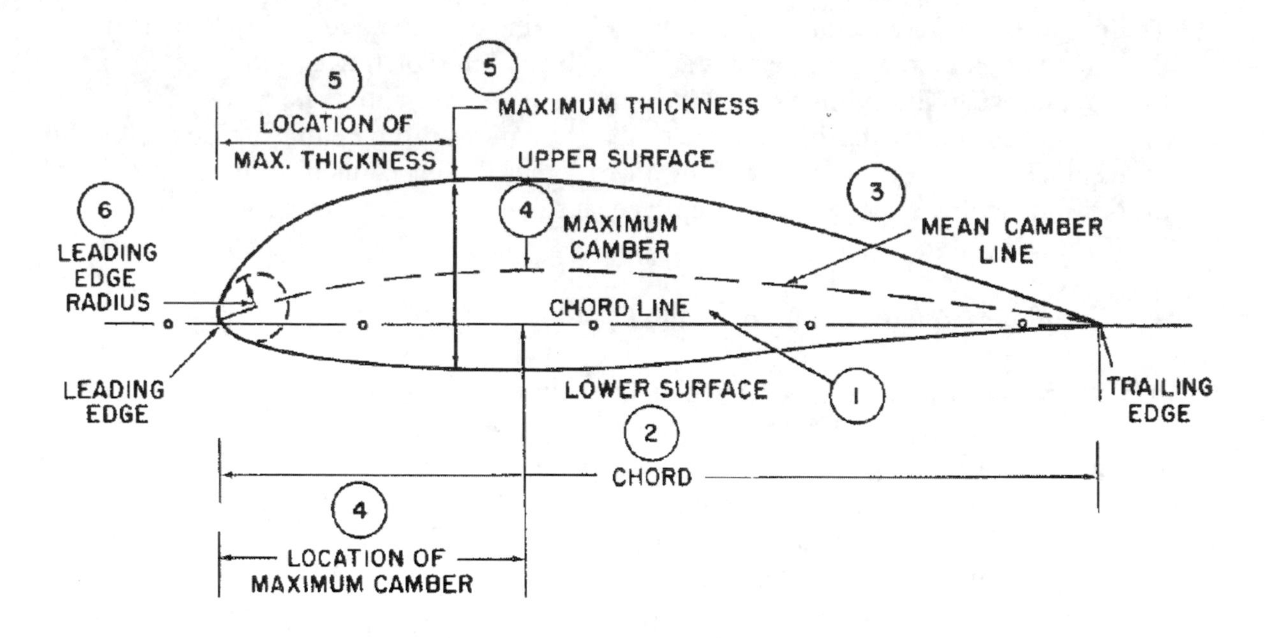

Figure 4.5. Cross Section of a Cambered Airfoil.
Source: Aerodynamics for Naval Aviators

Different airfoils have different flight characteristics. Many thousands of airfoils have been tested in wind tunnels and in actual flight, but no one airfoil has been found that satisfies every flight requirement. The weight, speed, and purpose of each aircraft dictate the shape of its airfoil. The most efficient airfoil for producing the greatest lift is one that has a concave, or "scooped out" lower surface. As a fixed design, this type of airfoil sacrifices too much speed while producing lift and is not suitable for high-speed flight. Advancements in engineering have made it possible for today's high-speed jets to take advantage of the concave airfoil's high lift characteristics. Leading edge (Kreuger) flaps and trailing edge (Fowler) flaps, when extended from the basic wing structure, literally change the airfoil shape into the classic concave form, thereby generating much greater lift during slow flight conditions.

On the other hand, an airfoil that is perfectly streamlined and offers little wind resistance sometimes does not have enough lifting power to take the airplane off the ground. Thus, modern airplanes have airfoils that strike a medium between extremes in design. The shape varies according to the needs of the airplane for which it is designed.

4.7.1 Low Pressure Above the Airfoil

In a wind tunnel or in flight, an airfoil is simply a streamlined object inserted into a moving stream of air. If the airfoil profile were in the shape of a teardrop, the speed and the pressure changes of the air passing over the top and bottom would be the same on both sides. But if the teardrop shaped airfoil were cut in half lengthwise, a form resembling the basic airfoil (wing) section would result. If the airfoil were then inclined so the airflow strikes it at an angle (**angle of attack [AOA]**), Figure 4.6, the air moving over the upper surface would be forced to move faster than the air moving along the bottom of the airfoil. This increased velocity reduces the pressure above the airfoil.

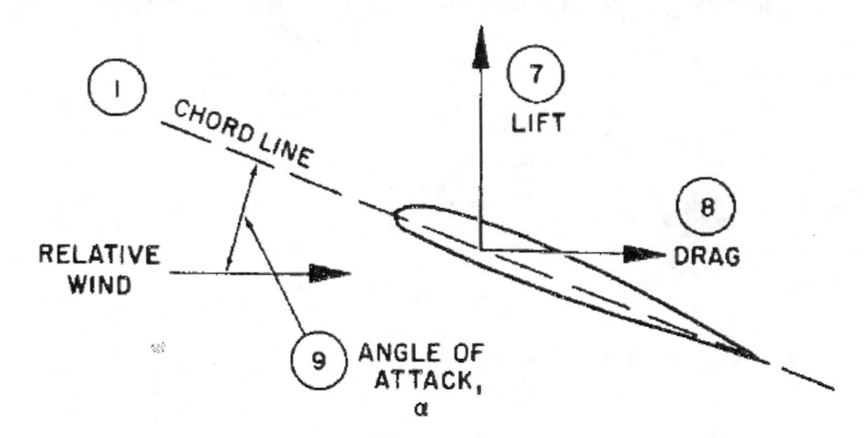

Figure 4.6. Cord Line, Relative Wind, and Angle of Attack.
Source: Aerodynamics for Naval Aviators

Applying Bernoulli's principle of pressure, the increase in the speed of the air across the top of an airfoil produces a drop in pressure. This lowered pressure is a component of total lift. The pressure difference between the upper and lower surface of a wing alone does not account for the total lift force produced.

The downward backward flow from the top surface of an airfoil creates a downwash. This downwash meets the flow from the bottom of the airfoil at the trailing edge. Applying Newton's third law, the reaction of this downward backward flow results in an upward forward force on the airfoil.

4.7.2 High Pressure Below the Airfoil

A certain amount of lift is generated by pressure conditions underneath the airfoil. Because of the manner in which air flows underneath the airfoil, a positive pressure results, particularly at higher angles of attack. But there is another aspect to this airflow that must be considered. At a point close to the leading edge, the airflow is virtually stopped (stagnation point) and then gradually increases speed. At some point near the trailing edge, it again reaches a velocity equal to that on the upper surface. In conformance with Bernoulli's principle, where the airflow was slowed beneath the airfoil, a positive upward pressure was created (i.e., as the fluid speed decreases, the pressure must increase). Since the pressure differential between the upper and lower surface of the airfoil increases, total lift increases. Both Bernoulli's principle and Newton's laws are in operation whenever lift is being generated by an airfoil.

4.7.3 Pressure Distribution

From experiments conducted on wind tunnel models and on full size airplanes, it has been determined that as air flows along the surface of a wing at different angles of attack, there are regions along the surface where the pressure is negative, or less than atmospheric, and regions where the pressure is positive, or greater than atmospheric. This negative pressure on the upper surface creates a relatively larger force on the wing than is caused by the positive pressure resulting from the air striking the lower wing surface. Figure 4.7 shows the pressure distribution along an airfoil at two different angles of attack

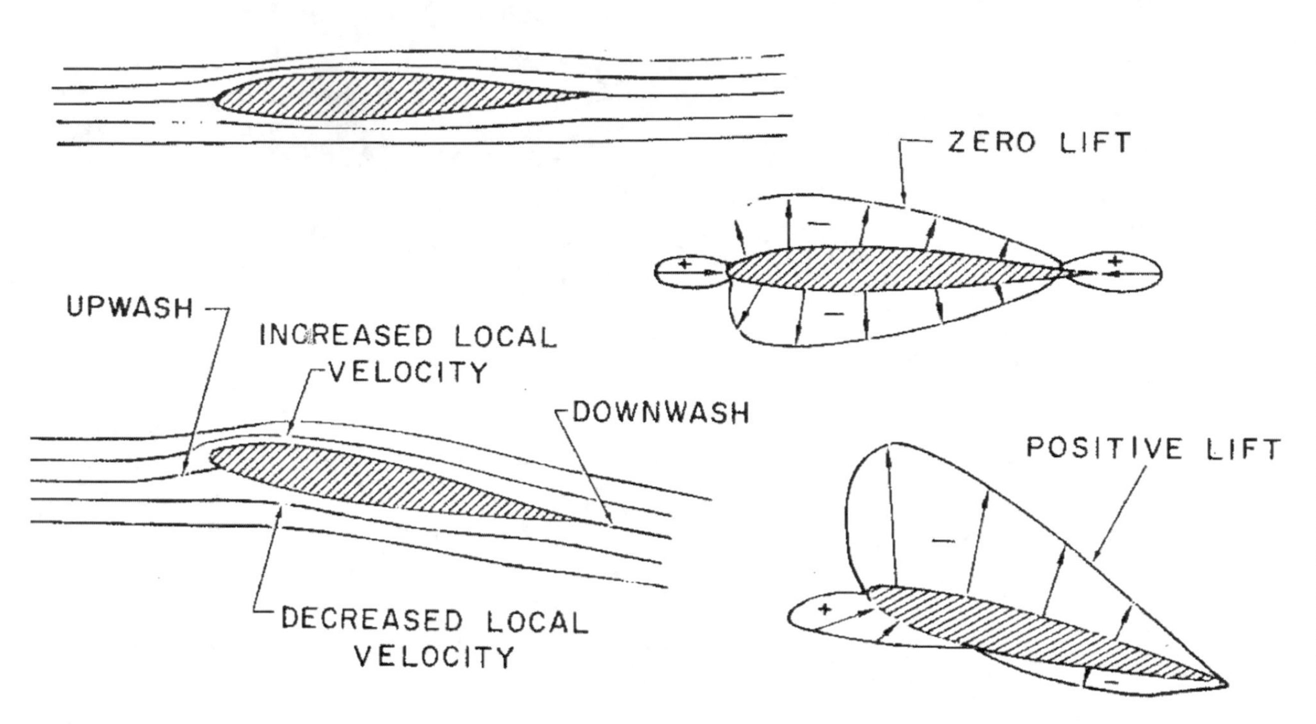

Figure 4.7. Streamlines and Pressure Distribution around an Airfoil.
Source: Aerodynamics for Naval Aviators

The velocity over the upper surface is greater than the velocity on the lower surface. This velocity increase can be visualized by realizing that the air must part and let the airfoil pass. Some of the air flows over the airfoil and some under the airfoil, but the airstreams above the airfoil travel at a much higher velocity as compared to the airstreams below the airfoil.

According to Bernoulli's equation, as velocity increases, dynamic pressure increases, causing static pressure to decrease. Thus, an airfoil reduces static pressure on the upper surface by increasing the velocity. The static pressure on the upper surface of the airfoil is less than the static pressure on the lower surface. It is significant that the static pressure on both surfaces can be less than atmospheric pressure and still produce lift. The important point here is the pressure differential which is developed across the airfoil.

Figure 4.7 shows the pressure differential by using vectors that point away from the surface of the airfoil to represent pressure below atmospheric, and using vectors that point toward the airfoil to represent pressures greater than atmospheric.

4.7.3.1 Aerodynamic Force

If pressure is applied to an area, a force is generated (Force = Pressure x Area). To obtain the net resulting pressure, the pressure differentials between the upper and lower surfaces are added algebraically. This net pressure is multiplied by the area of the airfoil to obtain the "aerodynamic force." This force acts at a point on the chord line called the "center of pressure." As the angle of attack is changed, the center of pressure moves back and forth along the chord. For the purposes of this discussion here, we consider the forces acting through a point called the "**aerodynamic center**," which is a stationary point 25 percent of the chord length aft of the leading edge (called **25 percent chord**). The significance of the aerodynamic center is that the pitching moment coefficient about this point does not vary with angle of attack. Thus, all wing forces may be resolved into one aerodynamic force, considered to act through the aerodynamic center. For positive cambered airfoils, this moment is nose down (negative). In this case of symmetrical airfoils, there is practically no movement of the center of pressure and the movement about the aerodynamic center is zero. The importance of the aerodynamic center will be realized when studying stability in later chapters.

The **aerodynamic force** is the total force acting on the airfoil, but with changes in the angle of attack and velocity, the direction and the magnitude of the aerodynamic force will change. For this reason, it becomes difficult to use the aerodynamic force to predict the performance of the aircraft and it somewhat complicates analysis of the airfoil's capabilities. *Breaking the aerodynamic force into two component forces in relation to the relative wind or flightpath simplifies analysis. The component of the aerodynamic force that is perpendicular to the relative wind is called lift, and the component of the aerodynamic force that acts parallel to the relative wind is called drag.* Note that these forces, lift and drag, are components of the aerodynamic force perpendicular and parallel to the relative wind, not the horizon. (Figure 4.6)

4.7.4 Lift Equation

We now have all of the variables to construct our lift equation. The air density (ρ) is primarily based upon temperature and pressure. The velocity (V) is squared and dependent on true air speed. This portion of the equation is the **dynamic pressure**. It is equal to the **indicated airspeed**.

The **coefficient of lift (C_L)** is dependent on the angle of attack and the shape of the airfoil. Finally, the wing planform area (S) completes our lift equation.

$$L = \frac{1}{2}\rho V^2 C_L S$$

L=Lift
ρ= Air density
V = Velocity
C_L= Coefficient of lift
S = Surface area

Chapter 5
Drag

The force which retards the forward motion of an aircraft through the air is referred to as drag. Drag acts parallel and opposite to the flightpath and relative wind. Since drag tends to retard motion and increase fuel consumption, performance objectives such as range, endurance, and maximum velocity are all affected by drag. In these instances, the pilot will fly with a "clean aircraft" (reducing the drag) as much as possible. Because drag requirements vary according to flight conditions the pilot must understand this force to obtain the required performance from the aircraft. There are two types of drag produced at subsonic speeds, parasite and induced drag. The first is called parasite because it does not aid flight. Induced drag is a result of an airfoil developing lift.

5.1 Parasite Drag

Parasite drag is composed of all the forces that work to slow an aircraft's movement. As the term parasite implies, it is the drag that is not associated with the production of lift. This includes the displacement of the air by the aircraft, turbulence generated in the airstream, or a hindrance of air moving over the surface of the aircraft and airfoil. There are many types of parasite drag. The most significant are: form drag, interference drag, and skin friction.

The formula for determining parasite drag is:

$$Dp = \frac{1}{2}\,\rho\,V^2\,C_{dp}\,S$$

Dp = Parasite drag

ρ = Air density

V = Velocity

C_{dp} = Coefficient of parasite drag

S = Surface area

5.1.1 Form Drag

Form drag is the portion of parasite drag generated by the aircraft due to its shape and airflow around it. Examples include the engine cowlings, antennas, and the aerodynamic shape of other components. When the air has to separate to move around a moving aircraft and its components, it eventually rejoins after passing the body. How quickly and smoothly it rejoins is representative of the resistance that it creates which requires additional force to overcome.

Figure 5.1. The Shape of an Object Affects Form Drag.
Source: Aerodynamics for Pilots, ATC Pamphlet 51-3

We can study the effect of shape on drag by comparing the values of drag coefficients for any two objects as long as the same reference area is used and the Mach number and Reynolds number (defined in Section 5.1.3 below) are matched. All of the drag coefficients on Figure 5.2 were produced in low speed (subsonic) wind tunnels and at similar Reynolds numbers, except for the sphere. A quick comparison shows that a flat plate gives the highest drag and a streamlined symmetric airfoil gives the lowest drag, by a factor of almost 30! *Shape has a very large effect on the amount of form drag produced.* The drag coefficient for a sphere is given with a range of values because the drag on a sphere is highly dependent on Reynolds number. Flow past a sphere, or cylinder, goes through a number of transitions with velocity. At very low velocity, a stable pair of vortices is formed on the downwind side. As velocity increases, the vortices become unstable and are alternately shed downstream. As velocity is increased even more, the boundary layer transitions to chaotic turbulent flow with vortices of many different scales being shed in a turbulent wake from the body. Each of these flow regimes produces a different amount of drag on the sphere. Comparing the flat plate and the prism, and the sphere and the bullet, we see that the downstream shape can be modified to reduce drag.

Figure 5.2. The Effect of Shape on Form Drag

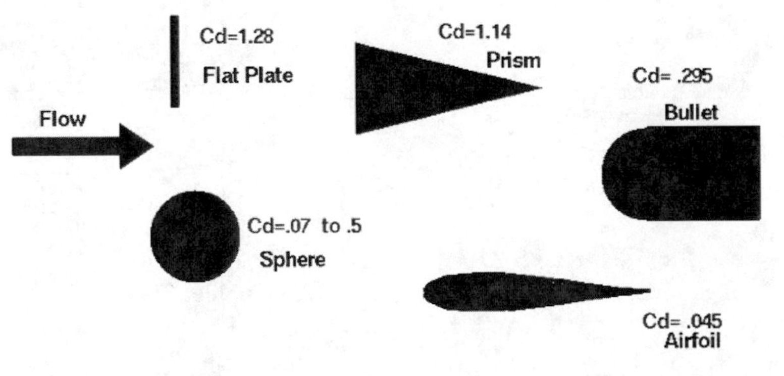

Source: NASA Beginner's Guide to Aerodynamics (BGA)

5.1.2 Interference Drag

Interference drag comes from the intersection of airstreams that creates eddy currents, turbulence, or restricts smooth airflow. Figure 5.3 shows the interference drag created at the juncture of two bodies. Figure 5.3 is an example where the intersection of the wing

and the fuselage at the wing root has significant interference drag. Air flowing around the fuselage collides with air flowing over the wing, merging into a current of air different from the two original currents. The most interference drag is observed *when two surfaces meet at acute and perpendicular angles*. **Fairings** are used to reduce this tendency. The wing and the fuselage junction is "faired" or "filled in" to reduce interference drag and to let the airstreams meet gradually instead of abruptly, thus reducing the turbulence formed. Due to the more critical pressure gradients on the upper wing surface, interference drag is most critical on the upper wing surface.
If wing tanks are hung on the wing of a fighter aircraft and the drag of the wing tank and the aircraft (separately) are known, the drag actually produced is higher than the sum of the drag of the individual components because of the interference drag created. This is a result of interference drag because the airstream must flow around the wing tank and the wing. Again, using fairings in the area between the wing and the wing tank, or increasing the distance between the two, will lower the effect of interference drag. Fairings and distance between lifting surfaces and external components (such as radar antennas hung from wings) reduce interference drag.

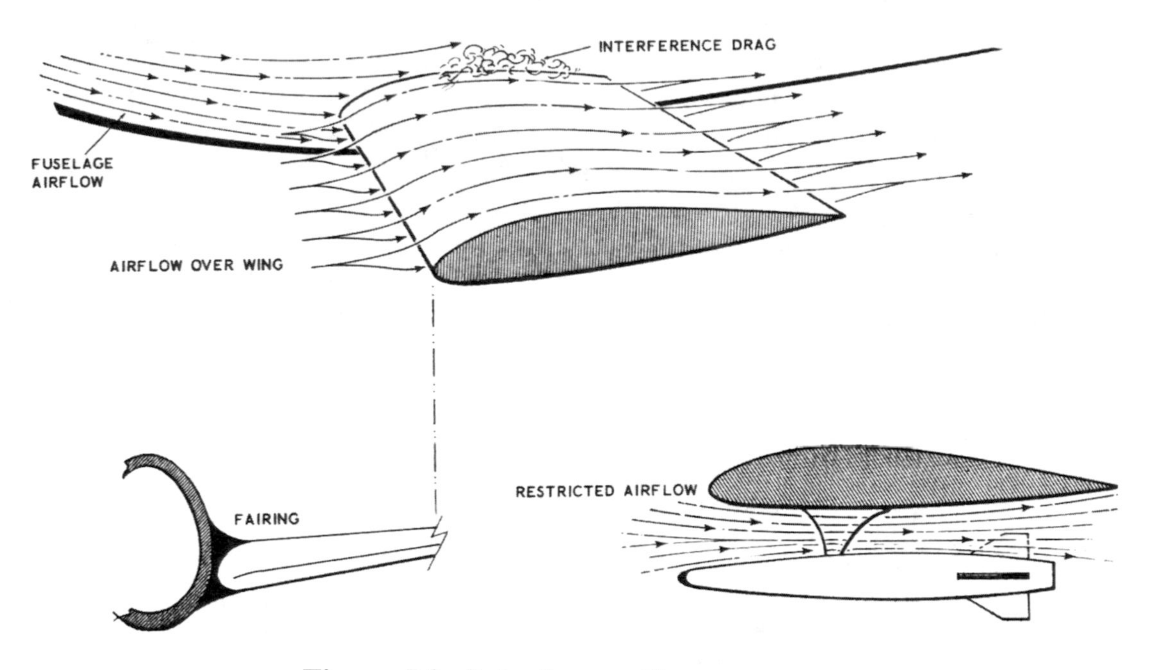

Figure 5.3. Interference Drag.
Source: Aerodynamics for Pilots, ATC Pamphlet 51-3

5.1.3 Skin Friction Drag

Covering the entire "wetted surface" of the aircraft is a thin layer of air called the **boundary layer**. The **wetted area** is the external area of the entire aircraft which is in contact with the external airflow. Figure 5.4 shows air flowing over a surface and the velocity profile of the air within the boundary layer. The air molecules on the surface have zero velocity in relation to the surface, but the layer above moves over the stagnant molecules on the surface because it is pulled along by the third layer. The velocities of the layers increase as distance from the surface is increased until free stream velocity is reached. The total distance between the surface and the point where the free stream velocity is reached is called the boundary layer and is about as thick at subsonic velocities as a playing card. The various layers of air within the boundary layer are sliding over one another and creating a force retarding motion, or a drag force due to the viscosity of the air. This type of drag force is called **skin friction drag** and is very small per square foot. However, when applied to large areas of such aircraft as transports, the force can become quite large and become a significant part of parasite drag.

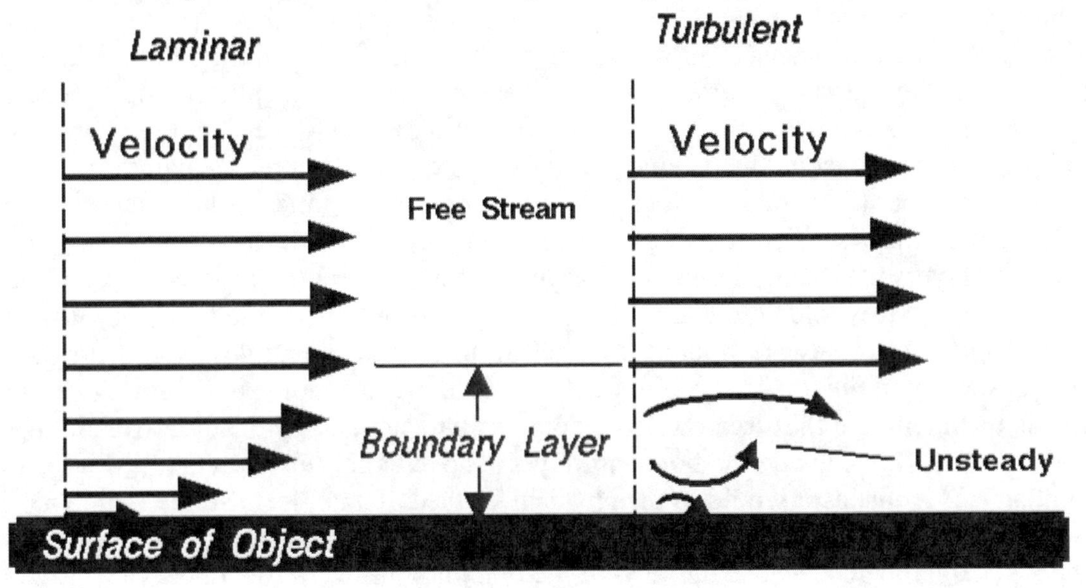

Velocity is zero at the surface (no – slip)

Figure 5.4. Boundary Layer.
Source: NASA Beginner's Guide to Aerodynamics (BGA)

In order to reduce the effect of skin friction drag, aircraft designers utilize flush mount rivets and remove any irregularities which may protrude above the wing surface. In addition, a smooth and glossy finish aids in transition of air across the surface of the wing. Since dirt on an aircraft disrupts the free flow of air and increases drag, keep the surfaces of an aircraft clean and waxed. Additionally, aluminum that is exposed to the atmosphere develops a coating of aluminum oxide which causes the surface to become rough and pitted which results in an appreciable resistance to smooth airflow. Paint can help, but it can oxidize, like aluminum, and increase the skin friction drag if it is left unattended. Therefore most large, high speed aircraft that cruise with parasite drag as the predominant drag have clean and highly polished surfaces to reduce the skin friction effects as much as possible.

As an object moves through a fluid, or as a fluid moves past an object, the molecules of the fluid near the object are disturbed and move around the object. Aerodynamic forces are generated between the fluid and the object. Two important properties of the air when determining skin friction drag are the **viscosity**, or stickiness, and the **compressibility**, or springiness, of the fluid. (Figure 5.7)

As the fluid moves past the object, the molecules right next to the surface stick to the surface. The molecules just above the surface are slowed down in their collisions with the molecules sticking to the surface. These molecules in turn slow down the flow just above them. The farther one moves away from the surface, the fewer the collisions affected by the object's surface. This creates a thin layer of fluid near the surface in which the

velocity changes from zero at the surface to the free stream value away from the surface. Engineers call this layer the *boundary layer* because it occurs on the boundary of the fluid.
Figure 5.5 shows the streamwise velocity variation from free stream to the surface. In reality, the effects are three dimensional. From the conservation of mass in three dimensions, a change in velocity in the streamwise direction causes a change in velocity in the other directions as well. There is a small component of velocity perpendicular to the surface which displaces or moves the flow above it. One can define the thickness of the boundary layer to be the amount of this displacement. The displacement thickness depends on the **Reynolds number** which is the ratio of inertial (resistant to change or motion) forces to viscous (heavy and gluey) forces and is given by the equation: Reynolds number (Re) equals velocity (V) times density (r) times a characteristic length (l) divided by the viscosity coefficient (mu). (Figure 5.7)
Boundary layers may be either laminar (layered), or turbulent (disordered) depending on the value of the Reynolds number. For lower Reynolds numbers, the boundary layer is laminar and the streamwise velocity changes uniformly as one moves away from the wall, as shown on the left side of Figure 5.5. For higher Reynolds numbers, the boundary layer is turbulent and the streamwise velocity is characterized by unsteady (changing with time) swirling flows inside the boundary layer. The external flow reacts to the edge of the boundary layer just as it would to the physical surface of an object. So the boundary layer gives any object an "effective" shape which is usually slightly different from the physical shape. This happens because the flow in the boundary has very low energy (relative to the free stream) and is more easily driven by changes in pressure. Flow separation is the reason for wing stall at high angles of attack. The effects of the boundary layer on lift are contained in the lift coefficient and the effects on drag are contained in the drag coefficient.

5.1.4 Cooling Drag and Leakage Drag

Cooling and leakage drag are the last two forms of parasite drag. **Cooling drag** refers to the air diverted into the engine compartment in reciprocating engines. (Engine) cooling drag has the characteristics of both form drag and skin drag. It is difficult to estimate and reduce.
Leakage drag refers to the air that is dumped out of the outflow valves in pressurized aircraft. Leakage drag is impossible to control and insignificant.

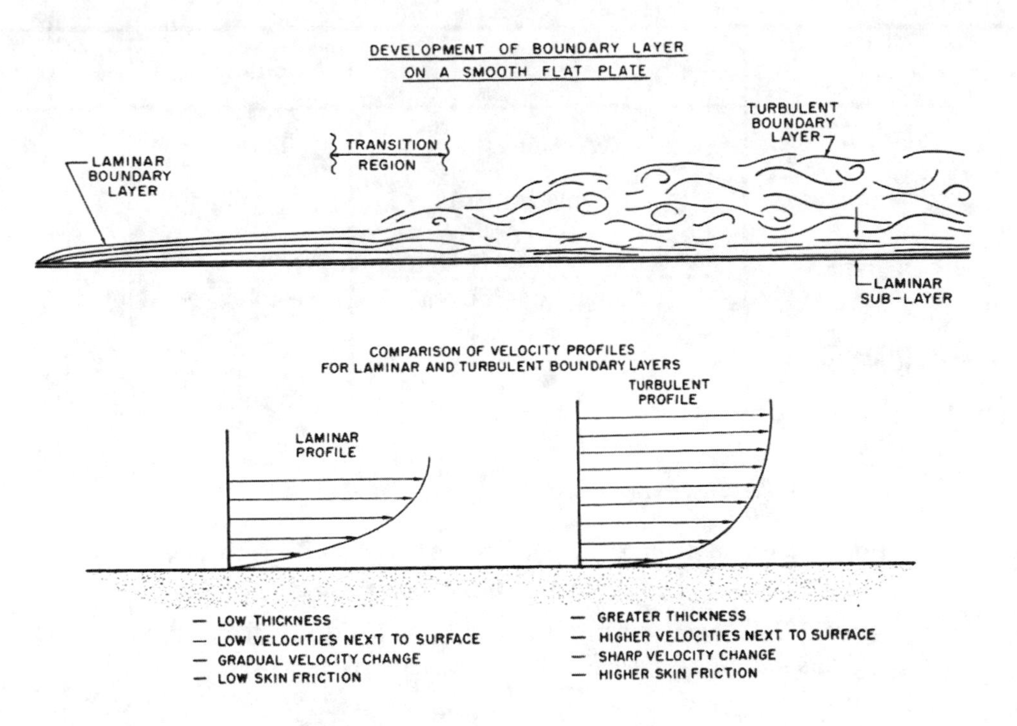

Figure 5.5. Boundary Layer Characteristics.
Source: Aerodynamics for Pilots, ATC Pamphlet 51-3

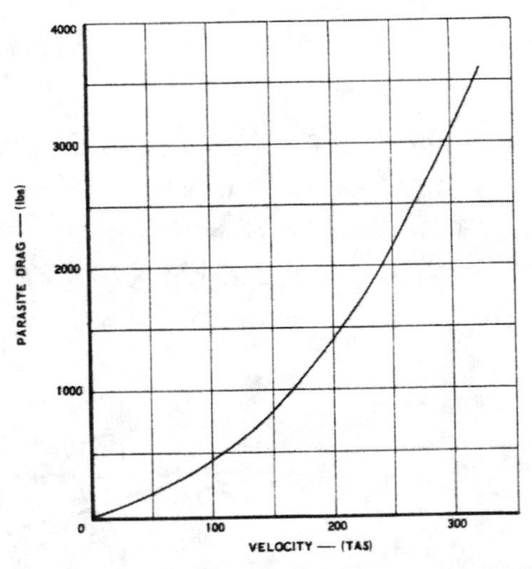

Figure 5.6. Parasite Drag Plotted Against Velocity.
Source: Aerodynamics for Pilots, ATC Pamphlet 51-3

	Viscosity	Compressibility
Characteristic	"Stickiness"	"Springiness"
Parameter	Reynolds (Re)	Mach (M)
Definition	density x velocity x length / viscosity coefficient	flow velocity / speed of sound
Equation	$\frac{\rho \times V \times L}{\mu}$	$\frac{V}{a}$

Aerodynamic Forces depend on Re and M

For a valid experiment, Reynolds Number and Mach Number must match flight conditions.

Figure 5.7. Aerodynamic Forces Depend upon Re and M.
Source: NASA Beginner's Guide to Aerodynamics (BGA)

5.2 Induced Drag

The portion of the total drag force that is due to the production of the lift force is called induced drag. This drag is induced as the wing develops lift.

5.2.1 Wing Circulation

When a wing is producing lift, there is a static pressure differential created across the wing with pressure well below atmospheric on most of the top surface and slightly below atmospheric on most of the bottom surface. This pressure differential induces a circulation flow to the top surface of the wing or to the low pressure area as shown in Figure 5.8.

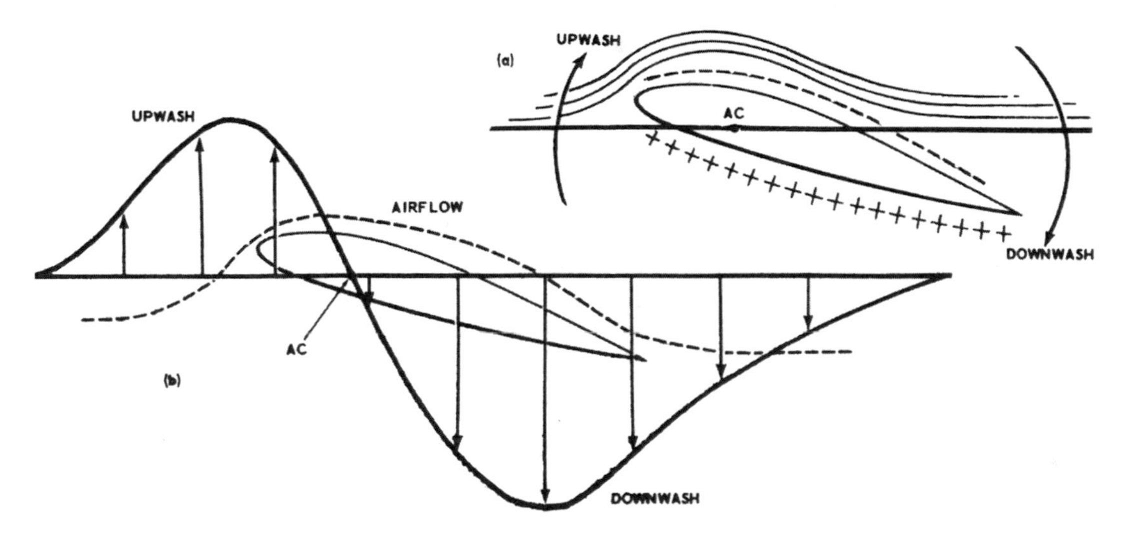

Figure 5.8. Bound Vortex.
Source: Aerodynamics for Pilots, ATC Pamphlet 51-3

This wing (Figure 5.8) has an infinite wing span so there are no wing tips for the air to flow equal to the downwash behind it, so these velocities cancel each other out. The circulation about the wing is called the **bound vortex**, and on this infinite wing there is no induced drag. Figure 5.8(b) shows the vicinity of the aerodynamic center. Note that there is no downwash at the aerodynamic center, the point where the aerodynamic force is generated. This is the reason that there is no induced drag in pure two-dimensional flow.

It would be desirable to design a wing without any induced drag (a wing with infinite wing span), but naturally a wing with an infinite span could not be constructed. *Not only does the air flow over the leading edge and create the circulation as mentioned above, but it also flows around and over the wingtips* as shown in Figure 5.11. As the wing moves through the air mass, the air trying to flow around the wingtip causes a vortex behind the wingtip. This wingtip vortex induces a spanwise flow and creates vortices all along the trailing edge of the wing. The trailing edge vortices are strongest at the tips and diminish in intensity progressing toward the center line of the wing. At the center line of the aircraft, there is no trailing-edge vortex because, in looking forward, the right wing vortices revolve counterclockwise and the left wing vortices revolve clockwise. The tip vortices cancel each other out at the center line.

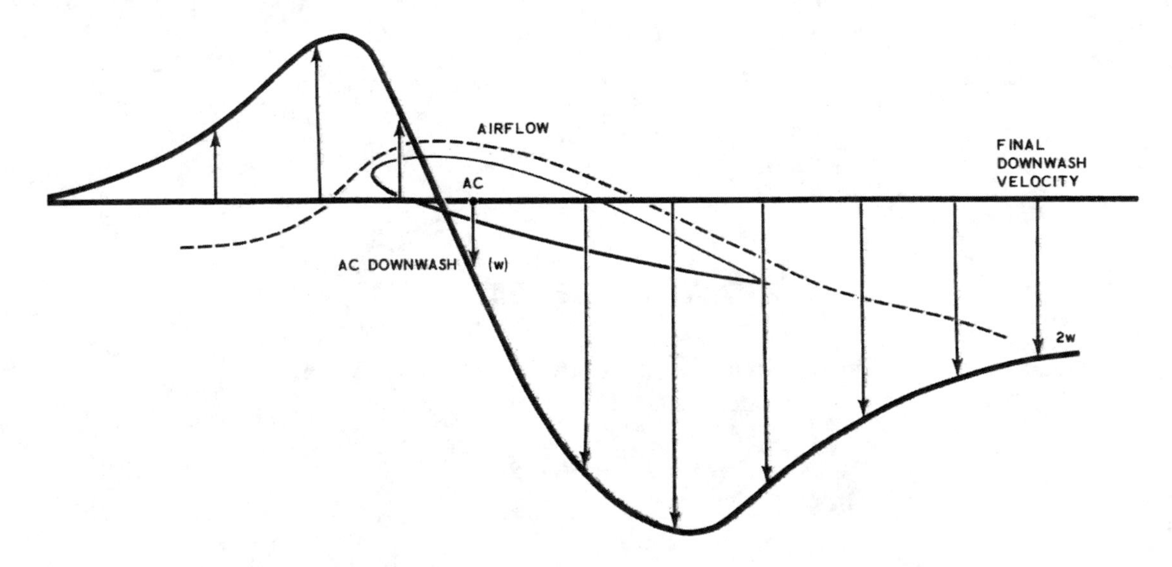

Figure 5.9. Vertical Velocity Vectors (Finite Wing).
Source: Aerodynamics for Pilots, ATC Pamphlet 51-3

The combination of the bound vortex and the trailing-edge vortices produces vertical velocities, as shown in Figure 5.9. In Figure 5.9 the dashed line shows the path of the air mass as it flows over the wing. Notice the downwash velocity at the aerodynamic center of the finite wing. This downwash vector added to the free stream relative wind vector results in another relative wind vector that is inclined to the actual flightpath. From wingtip to wingtip, the magnitude of the downwash vector varies as the intensity of the trailing-edge vortices varies.

Figure 5.10 shows the vector diagram using an average downwash velocity vector. The downwash vector (w) has been reversed and added to the opposite end of the free stream relative wind vector to simplify the diagram. The average relative wind, being the flow that actually affects the wing, is inclined to the free airstream relative wind at an angle of

α_i, the induced angle of attack. The force that is produced by the wing (labeled "L" for "lift" on the diagram) is perpendicular to the average relative wind. The lift force that is perpendicular to the free stream relative wind is named the **effective lift**. The component of the effective lift that is parallel to the free stream relative wind is the induced drag (D_i).

Lift is perpendicular to the average relative wind and the effective lift is perpendicular to the remote free stream relative wind.

5.2.2 Angle of Attack vs. Induced Drag

In examining the induced drag, note that if the angle of attack is increased, the pressure differential increases. This increases the downwash, which increases the induced angle of attack. The result is a greater angle between the lift and resultant force vectors and, therefore, an increase in the induced drag.

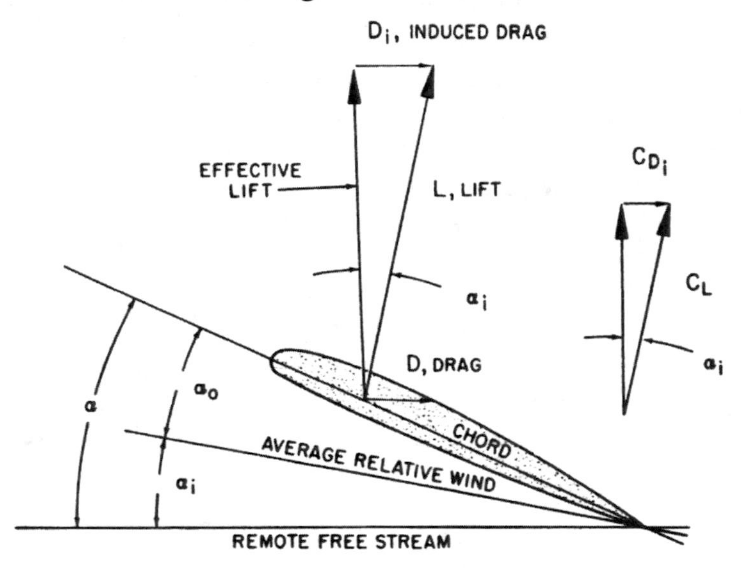

Figure 5.10. Induced Drag Vector Diagram.
Source: Aerodynamics for Pilots, ATC Pamphlet 51-3

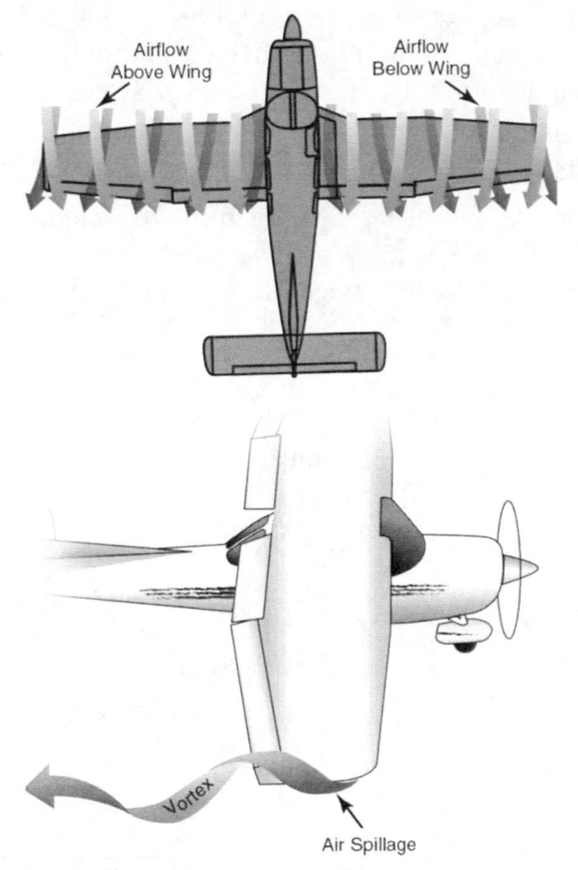

Figure 5.11. The Creation of Vortices on a Finite Wing.
Source: Pilot's Handbook of Aeronautical Knowledge

5.2.3 Induced Drag Equation

The induced drag is directly related to the angle of attack, and therefore to the coefficient of lift. The **induced drag equation** can be derived from the aerodynamic force equation and appears as:

$$D_i = \frac{1}{2}\rho V^2 S\, c_{D_i}$$

This is the aerodynamic force equation with a new coefficient, c_{D_i}. The **induced drag coefficient** can be found from the following equation:

$$c_{D_i} = \frac{C_L^2}{\pi\ AR\ \epsilon}$$

This equation is not derived in this text, but can be found in most aeronautical engineering textbooks. The AR is the **aspect ratio** and is equal to the wing span squared divided by the wing area ($AR = b^2/s$). This general equation is valid for any wing. If the wing is rectangular, the aspect ratio is the ratio of the span to the chord ($AR = b/c$)
The **span efficiency factor** of the wing is shown by the Greek letter Epsilon ϵ. For subsonic flight, it may be defined as the ratio between the theoretical minimum drag-due-to-lift to the actual drag-due-to-lift applicable to the particular wing, as derived from flight test. An efficiency factor of one represents the elliptical planform. A swept wing aircraft develops its best efficiency factor at cruise airspeeds. At lower airspeeds (higher

angle of attack), flight tests and wind tunnel experiments show that ϵ decreases. This factor and the lower aspect ratio of high performance aircraft (as compared to conventional straight wings of low speed aircraft) cause swept wing airplanes to have more induced drag at low speeds.

Although the induced drag equation $D_i = \frac{1}{2} \rho V^2 S \, c_{D_i}$ makes induced drag appear to be directly proportional to V^2, this is *not* actually true. As velocity is increased, the pilot must decrease the angle of attack to maintain level flight, thereby decreasing induced drag. To understand the relationship between velocity and induced drag, solve the lift equation for C_L.

$$C_L = \frac{2L}{\rho V^2 S}$$

Then square both sides of the above equation and divide by AR ϵ to get:

$$\frac{C_L^2}{\pi \, AR \, \epsilon} = \left(\frac{2L}{\rho \, V^2 \, S}\right)^2 \left(\frac{1}{\pi \, AR \, \epsilon}\right)$$

From earlier in paragraph 5.2.3 we stated that $c_{D_i} = \frac{C_L^2}{\pi \, AR \, \epsilon}$ so,

$$c_{D_i} = \left(\frac{2L}{\rho \, V^2 \, S}\right)^2 \left(\frac{1}{\pi \, AR \, \epsilon}\right)$$

Substituting $\frac{C_L^2}{\pi \, AR \, \epsilon} = \left(\frac{2L}{\rho \, V^2 \, S}\right)^2 \left(\frac{1}{\pi \, AR \, \epsilon}\right)$ into $D_i = \frac{1}{2} \rho V^2 S \, c_{D_i}$ results in:

$$D_i = \left(\frac{\rho \, V^2 S}{2}\right) \left(\frac{4L^2}{\rho^2 \, V^4 \, S^2 \, \pi \, AR \, \epsilon}\right)$$

$$D_i = \left(\frac{2L^2}{\rho \, V^2 \, S \, \pi \, AR \, \epsilon}\right)$$

$$D_i = \left(\frac{1}{V^2}\right) \left(\frac{2L^2}{\rho \, S \, \pi \, AR \, \epsilon}\right)$$

Substituting $AR = \frac{b^2}{S}$ results in:

$$D_i = \left(\frac{1}{V^2}\right) \left(\frac{2L^2}{\rho \, \pi \, b^2 \, \epsilon}\right)$$

For level flight we will assume that lift (L) equals weight (W). This results in the final equation for induced drag:

$$\mathbf{D_i = \frac{2W^2}{V^2 \rho \, \pi \, b^2 \, \epsilon}}$$

5.2.3.1 Factors that Affect Induced Drag

By changing the variables in the equation for induced drag one at a time you can see their effects on the induced drag. Note that the induced drag is directly proportional to the square of the weight. Assuming level 1-g flight, weight will equal lift. Induced drag is the drag produced due to the production of lift.

Drag - Pounds

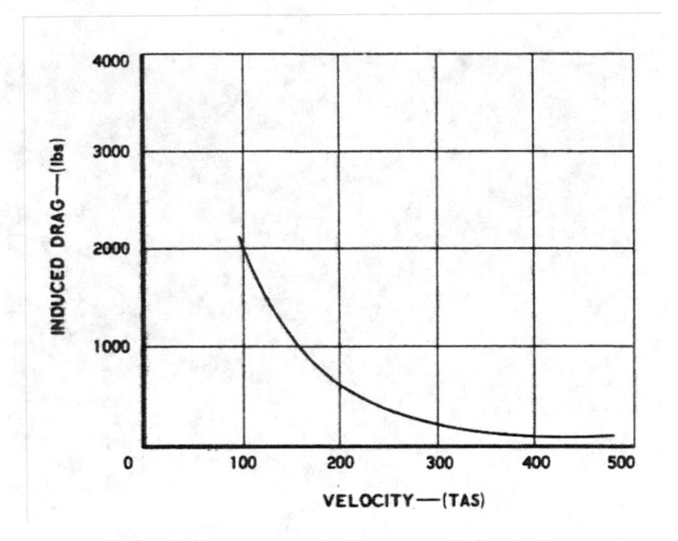

Figure 5.12. Induced Drag Plotted Against Velocity.
Source: Aerodynamics for Pilots, ATC Pamphlet 51-3

Referring to the equation for induced drag, the most significant factors are weight, velocity, and wing span. Doubling the weight causes the induced drag to increase by a factor of four. Additionally, induced drag decreases with speed, varying inversely as the square of the velocity. Figure 5.12 shows how induced drag changes with respect to velocity when all other factors are held constant. The wing span also has a large effect on induced drag. Induced drag decreases with wing span, varying inversely as the square of the
wing span. The longer the wing span, the lower the induced drag.

Other factors affecting induced drag to a smaller degree are density (ρ) and the span efficiency factor (ϵ). Induced drag decreases with greater air density. We will increase density (and decrease induced drag) by increasing the pressure, lowering the temperature, or lowering the altitude. Figure 5.10 shows how induced drag changes as velocity changes. In Figure 5.10 all other factors such as weight, density, wing span, and wing efficiency are held constant.
Additionally, increasing the span efficiency factor (ϵ) will decrease induced drag. The elliptical wing has the highest span efficiency factor of 1.0. Other wings are less efficient and have span efficiency factors of .85 to .95. Hence, the more efficient the wing, the less induced drag produced.

5.3 Total Drag

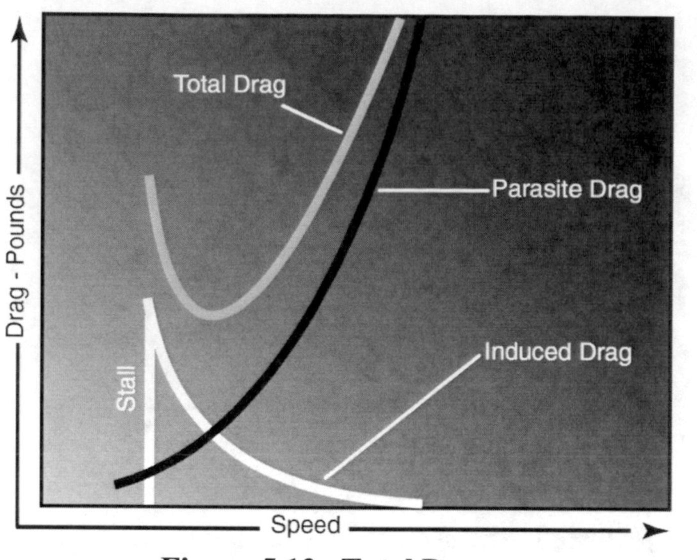

Figure 5.13. Total Drag.
Source: Pilot's Handbook of Aeronautical Knowledge.

Total drag is the sum of the induced drag (drag due to the production of lift) and the parasite drag (drag from all other sources) and results in a curve which is characterized by a decrease and then an increase in drag as the velocity of the airplane increases (assuming level flight, constant weight, configuration, and altitude). Figure 5.13 shows induced, parasite, and total drag plotted against velocity. There are other factors that affect total drag to a large degree. These are weight, configuration, and altitude.

5.3.1 Weight Affects Total Drag

Figure 5.14 reflects the changes in the total drag curve resulting from a 50 percent increase in the gross weight of the aircraft. The most obvious difference between the two curves is the large separation at the lower velocities. Since induced drag is predominant at these velocities where the angle of attack is relatively high, the increase in the gross weight of an aircraft must influence induced drag. At higher velocities, where parasite drag is predominant, there is little difference between the performance curves because parasite drag is not affected by a change in weight.

The effect of a change in weight on induced drag is explained very easily. Assuming lift is equal to weight, an increase in weight requires the same increase in lift. If this is to be accomplished at a constant velocity, then C_L must be increased. Since the induced drag is directly proportional to the C_L squared, this increase in the angle of attack increases the induced drag at the velocity in question.

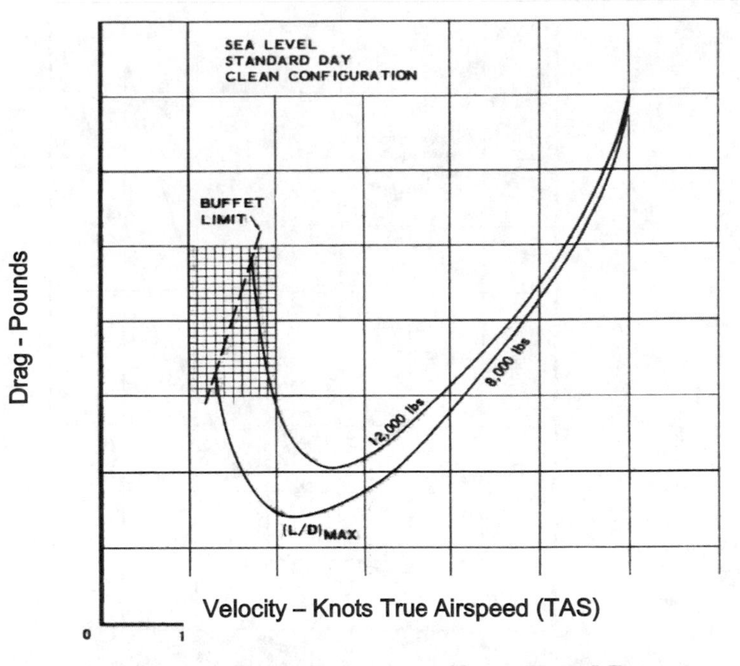

Figure 5.14. Change in Weight Affects Total Drag.
Source: Aerodynamics for Pilots, ATC Pamphlet 51-3

5.3.2 Configuration Changes Affect Total Drag

A change in the configuration of an aircraft does not have the same effect on the performance curves as does a change in the weight. Changing the configuration of the aircraft can be accomplished by lowering the landing gear, extending speed brakes, lowering the flaps, and any other method that affects the parasite drag on the airplane. A change in configuration changes the equivalent parasite area. This area multiplied by the dynamic pressure (q) of the airstream is the parasite drag. This is reflected in Figure 5.15 where the equivalent parasite area of a hypothetical aircraft has been increased 50 percent (assuming no change in weight).

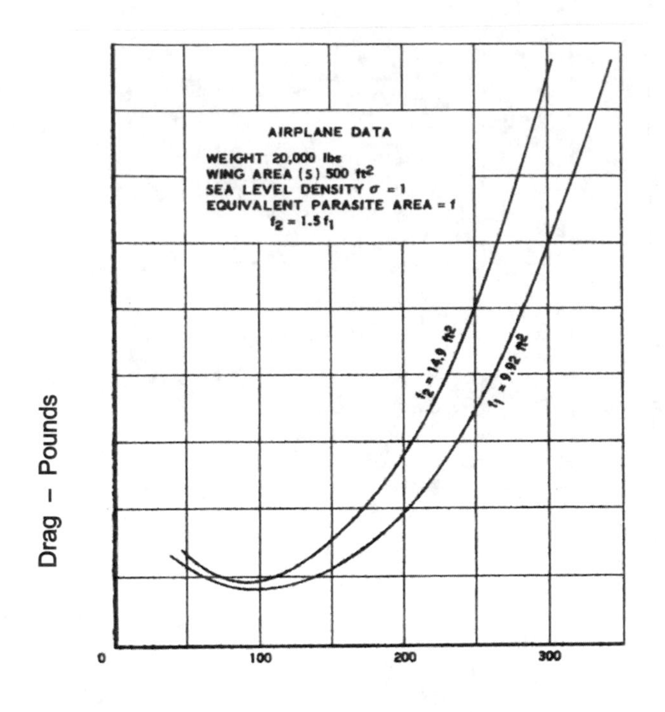

Velocity – Knots TAS

Figure 5.15 Configuration Affects Total Drag.

Since changes in the configuration of the aircraft produce changes only in the parasite drag, the performance curves show the greatest variations at the higher velocities where parasite drag is predominant. Naturally, the increased drag on the aircraft requires more power or thrust to maintain a certain velocity.

5.3.3 Altitude Changes Affect Total Drag

Figure 5.16 shows how an increase in altitude results in a decrease in the air density. This means that the aircraft may increase its velocity (maintaining C_D constant) without increasing the drag force. A hypothetical aircraft can experience the same drag at 345 knots at 20,000 feet that it did at 240 knots at sea level. Because the drag curve moves to the right as altitude increases, the drag to maintain a lower velocity increases.

With a true airspeed (TAS) of 220 knots at sea level the aircraft experiences 875 lbs of total drag; but at 20,000 feet, drag increases to 1,400 lbs at the same velocity. This is the result of induced drag. To maintain the same airspeed at a higher altitude, the angle of attack must be increased because of the reduction in air density. This results in an increase in induced drag.

At the higher velocities, there is a definite reduction in the total drag. At 400 knots, the thrust required to maintain a speed decreases from 1,550 lbs to almost 950 lbs as the altitude is increased from sea level to 20,000 feet. This shows the important effects of an increase in altitude on the total drag curve. The drag remains constant, but the velocity is higher.

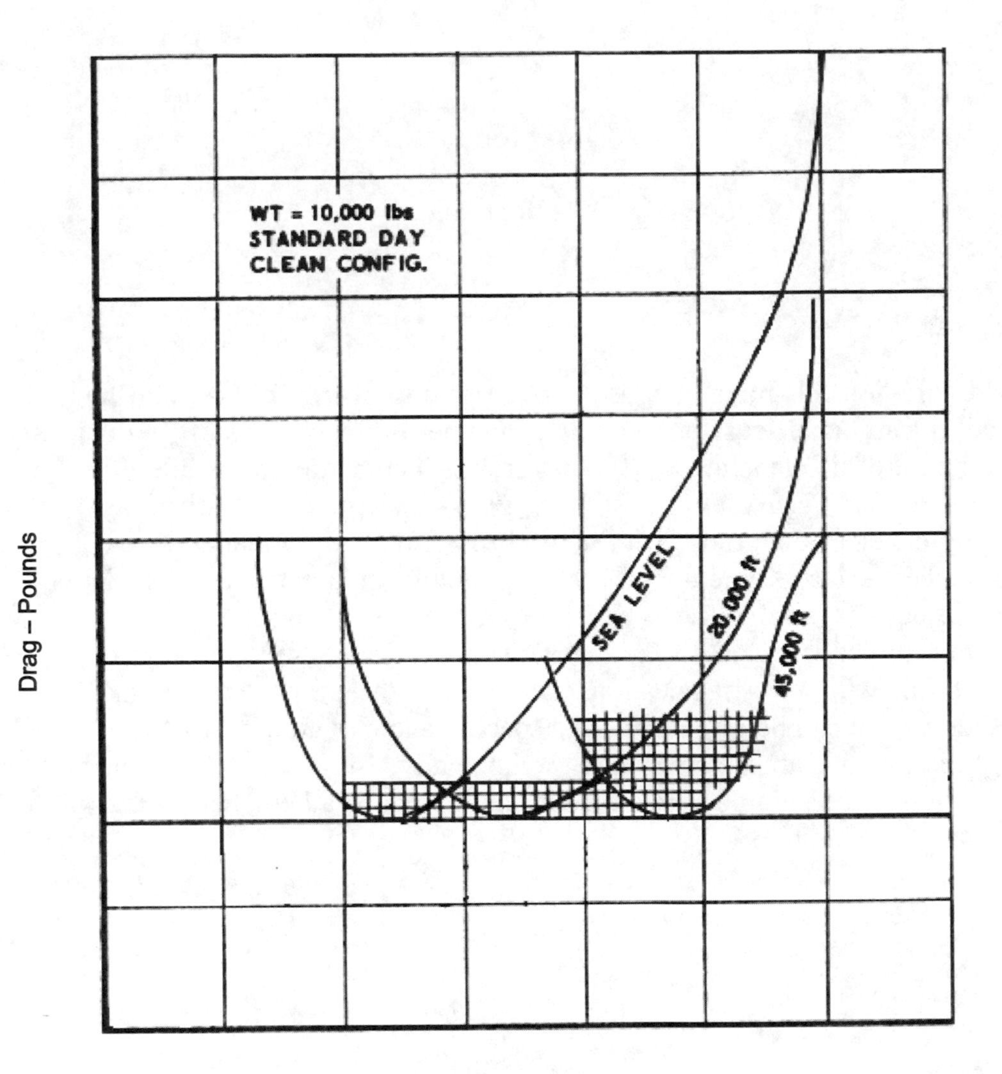

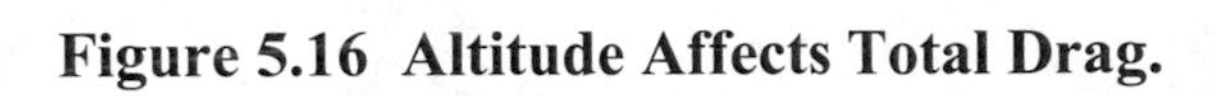

Figure 5.16 Altitude Affects Total Drag.

Chapter 6

Power and Performance

6.1 Power and Performance

As discussed in previous chapters, drag is the price paid to obtain lift. A short review of drag is needed to lead our discussion into power and performance. The **lift to drag ratio (L/D)** is the amount of lift generated by a wing or airfoil compared to its drag. A ratio of L/D indicates **airfoil efficiency**. Aircraft with higher L/D ratios are more aerodynamically efficient than those with lower L/D ratios. An airplane has a high L/D ratio if it produces a large amount of lift or a small amount of drag. Under cruise conditions lift is equal to weight. A high lift aircraft can carry a large payload. Under cruise conditions thrust is equal to drag. A low drag aircraft requires low thrust. In unaccelerated flight with the lift and drag data steady, the proportions of the C_L and coefficient of drag (C_D) can be calculated for a specific angle of attack (AOA). Typically lift and drag coefficients are determined experimentally using a wind tunnel, but for simple explanation and examples they can be determined mathematically and plotted graphically as done in Figure 6.1.

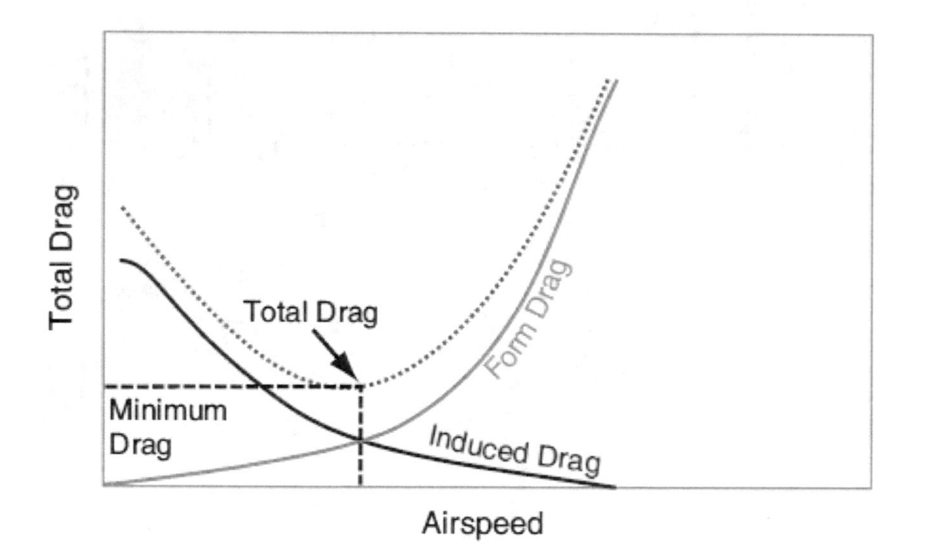

Figure 6.1. Drag versus Speed.
Source: Pilot's Handbook of Aeronautical Knowledge

The L/D ratio is determined by dividing the C_L by the C_D, which is the same as dividing the lift equation by the drag equation. All variables except coefficients cancel out.

L = Lift in pounds
D = Drag

Where L is the lift force in pounds, C_L is the lift coefficient, ρ is density expressed in slugs per cubic feet, V is velocity in feet per second, q is dynamic pressure per square feet, and S is the wing area in square feet.

C_D is the ratio of drag pressure to dynamic pressure. Typically at low angles of attack, the drag coefficient is low and small changes in angle of attack create only slight changes in the drag coefficient. At high angles of attack, small changes in the angle of attack cause significant changes in drag.

$$L = \frac{C_L \,.\, \rho \,.\, V^2 \,.\, S}{2}$$

$$D = \frac{C_D \,.\, \rho \,.\, V^2 \,.\, S}{2}$$

The above formulas represent the coefficient of lift (C_L) and the coefficient of drag (C_D) respectively. The shape of an airfoil and other lift-producing devices (e.g., flaps) affect the production of lift and alter with changes in the AOA. The lift/drag ratio is used to express the relation between lift and drag and is determined by dividing the lift coefficient by the drag coefficient—C_L/C_D.

Note in Figure 6.2 that the lift curve reaches its maximum for this particular wing section at 20° AOA, and then rapidly decreases. The drag curve (yellow) increases very rapidly from 14° AOA and completely overcomes the lift curve at 21° AOA. The lift/drag ratio (green) reaches its maximum at 6° AOA, meaning that at this angle, the most lift is obtained for the least amount of drag.

Note that the maximum lift/drag ratio (L/D_{MAX}) occurs at one specific C_L and AOA. If the aircraft is operated in steady flight at L/D_{MAX}, the total drag is at a minimum. Any AOA other than that associated with L/D max will result in more total drag. Figure 6.2 depicts the L/D_{MAX} by the lowest portion of the orange line labeled "total drag." The configuration of an aircraft has a large impact on the L/D.

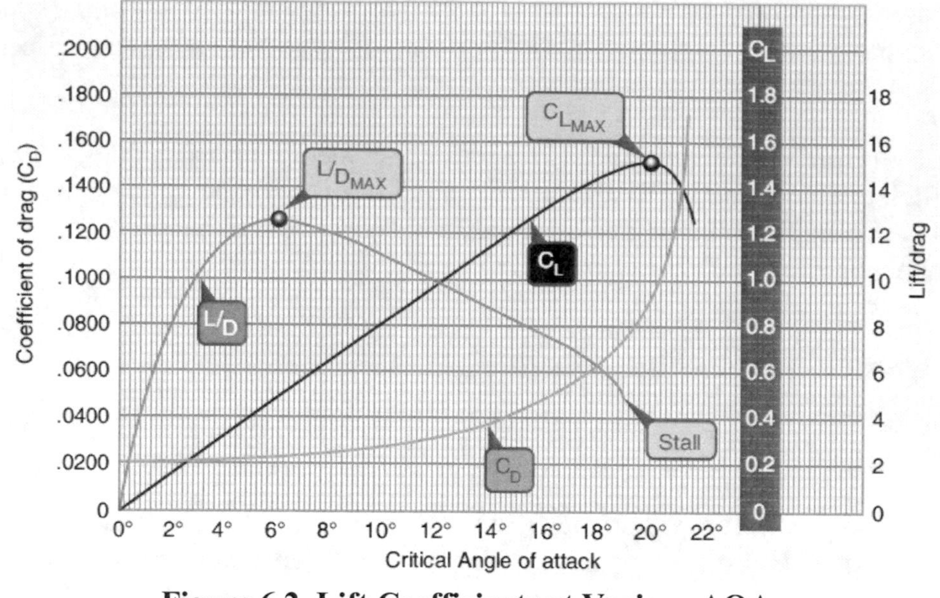

Figure 6.2. Lift Coefficients at Various AOA.
Source: Pilot's Handbook of Aeronautical Knowledge

While the parasite drag predominates at high speed, induced drag predominates at low speed. For example, if an aircraft in a steady state flight condition at 100 knots is then accelerated to 200 knots, the parasite drag becomes four times greater, but the power required to offset the additional induced drag is eight times greater. Conversely, when the aircraft is operated in steady state level flight at twice as great a speed, the induced drag is one-fourth the original value, and the power required to overcome that drag is only one-half the original value.

This is an example of why putting a bigger engine on an existing airframe significantly improves cruise performance. The new engine adds extra weight, drag increases exponentially with speed, and even more power is required to overcome the increase in both parasite and induced drag.

6.1.1 Thrust to Weight Ratio

Just as the lift to drag ratio is an efficiency parameter for total aircraft aerodynamics, the **thrust to weight ratio** is an efficiency factor for total aircraft propulsion. From Newton's second law of motion_for constant mass, force **F** is equal to mass **m** times acceleration **a**:

$$F = m * a$$

If we consider a horizontal acceleration and neglect the drag, the net external force is the thrust **F**. From the Newtonian weight equation:

$$W = m * g$$

Where **W** is the weight and **g** is the gravitational constant equal to 32.2 ft/sec^2. Solving for the mass:

$$m = W / g$$

And substituting in the force equation:

$$F = W * a / g$$
$$F / W = a / g$$

F/W is the thrust to weight ratio and it is directly proportional to the acceleration of the aircraft. An aircraft with a high thrust to weight ratio is capable of high acceleration. For most flight conditions, an aircraft with a high thrust to weight ratio will also have a large quantity of excess thrust. High excess thrust results in a high rate of climb. If the thrust to weight ratio is greater than one and the drag is small, the aircraft can accelerate straight up like a rocket. Similarly, rockets must develop thrust greater than the weight of the rocket in order to lift off.

The propulsion system of an aircraft must perform two important roles:

- During cruise, the engine must provide enough thrust to balance the aircraft drag while using as little fuel as possible.
- During takeoff and maneuvering, the engine must provide additional thrust to accelerate the aircraft.

Thrust **T** and drag **D** are vector forces and are vector quantities which have a magnitude and a direction associated with them. The thrust minus the drag of the aircraft is called the **excess thrust** and is also a vector quantity. Considering Newton's second law of motion mass **m** times acceleration **a** is equal to the net external force **F** on an object:

$$F = m * a$$

For an aircraft, the horizontal net external force **Fh** is the excess thrust **Fex**.

$$Fex = Fh = T - D = m * a$$

Therefore, the acceleration of an aircraft is equal to the excess thrust divided by the mass of the aircraft.

$$a = (T - D) / m$$

The thrust divided by the mass of the aircraft is closely related to the thrust to weight ratio. Airplanes with high excess thrust, like fighter planes, can accelerate faster than aircraft with low excess thrust.

6.1.2 Steady State Flight

For the airplane to remain in steady level flight, equilibrium must be obtained by a lift equal to the airplane weight and a powerplant thrust equal to the airplane drag. Thus, the airplane drag defines the thrust required to maintain steady state level flight as seen in Figure 6.3

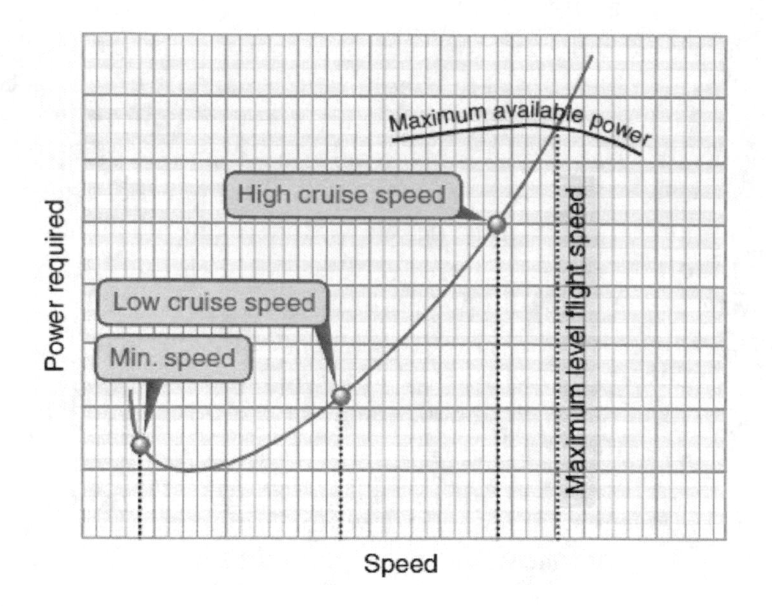

Figure 6.3. Power versus Speed.
Source: Pilot's Handbook of Aeronautical Knowledge

6.2 Basic Power Concepts

Aircraft aerodynamics involves the interaction of the four forces: lift, weight, thrust, and drag. In the previous section we discussed how the airplane's total drag determines the thrust required. The first basic issue we face is the difference between propeller-driven aircraft power and jet engine thrust. Power is what a propeller driven engine produces; thrust is what a jet engine produces. The propeller of the aircraft is said to produce thrust, not the engine. The thrust on a propeller driven aircraft decreases with an increase in velocity, in a jet aircraft thrust remains relatively constant with an increase in aircraft velocity—Figure 6.4.

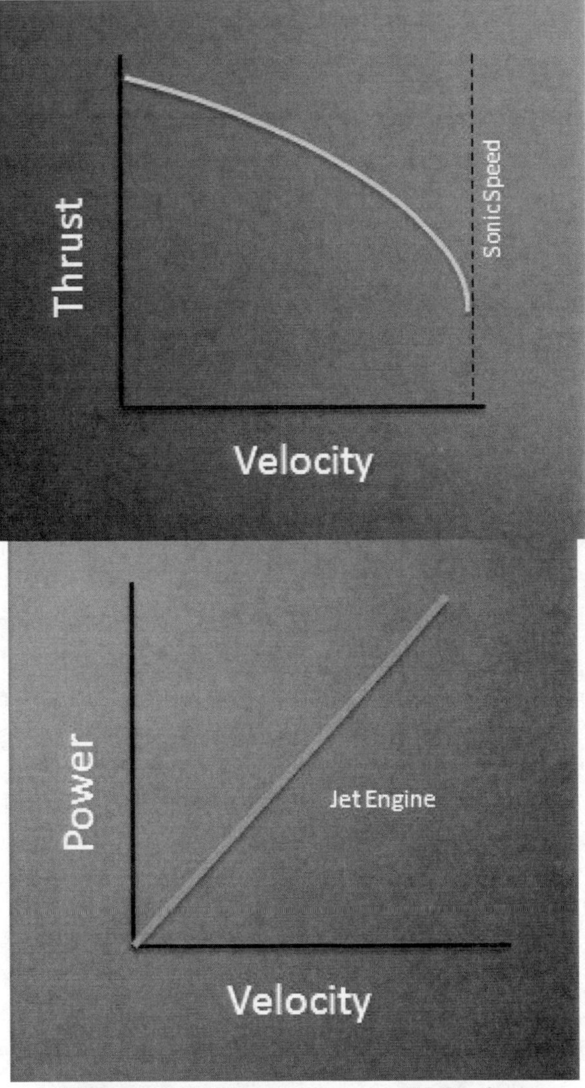

Figure 6.4. Thrust versus Velocity.
Source: Shelby Balogh, UND Aerospace

Therefore, the power required curve versus the power available curve for a propeller-driven aircraft and a jet aircraft will look different—Figure 6.6.

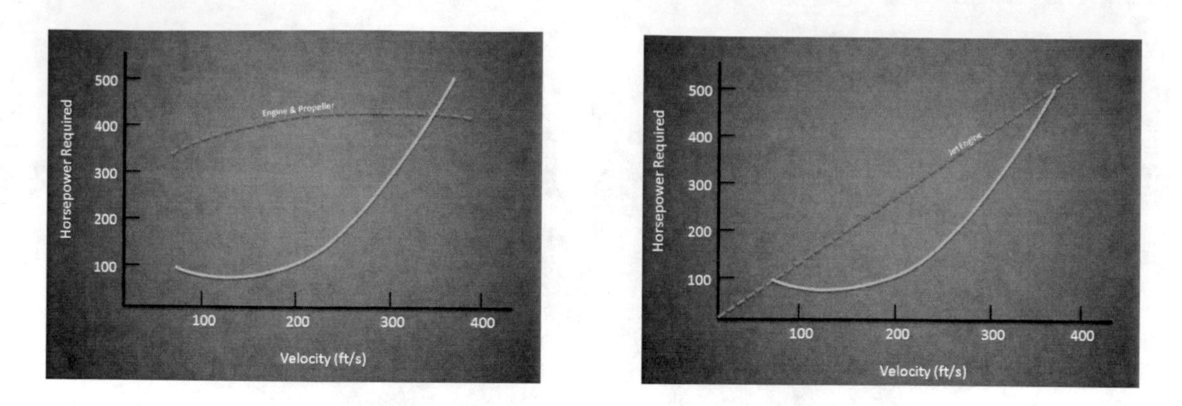

Figure 6.6 Power Required versus Power Available.
Source: Shelby Balogh, UND Aerospace

6.2.1 Propeller Efficiency

Propeller efficiency is a measure of how much power is absorbed (transmitted) by the propeller and turned into thrust. To begin to understand propeller efficiency one must start with a basic review of propeller principles. Propellers on aircraft consist of two or more blades and a hub. The blades are attached to the hub; the hub is attached to the crankshaft on a piston power aircraft and a gear reduction box on most turbo-prop aircraft. The propeller is simply a rotating wing that produces lift along the vertical axis. We call this lift force thrust—Figure 6.7.

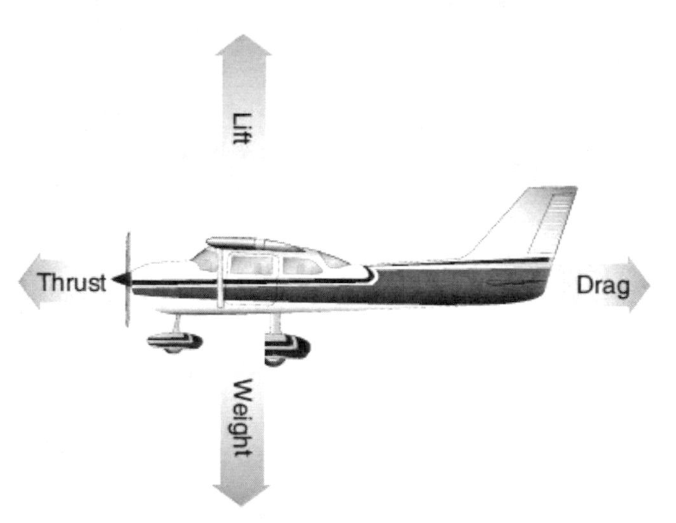

Figure 6.7. Forces in Flight.
Source: Pilot's Handbook of Aeronautical Knowledge

Looking at a cross section of the propeller blade, we can see it is similar to a cross section of an aircraft wing—Figure 6.8. The top portion of the blade is cambered like the top surface of the wing. The bottom portion is flat like the bottom surface of the wing.

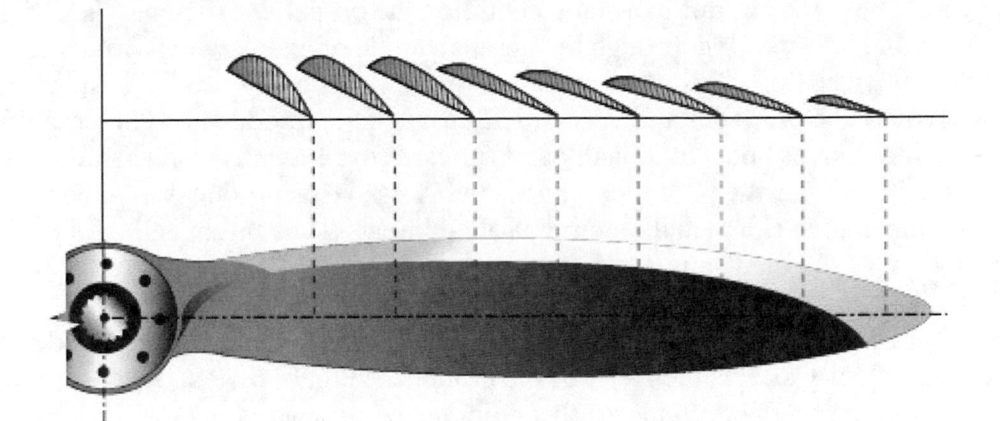

Figure 6.8. Propeller Cross Section.
Source: Pilot's Handbook of Aeronautical Knowledge

The chord line is an imaginary line drawn from the leading edge of the propeller blade to the trailing edge of the propeller blade. Blade angle, measured in degrees, is the angle between the chord of the blade and the plane of rotation—Figure 6.9. The pitch of the propeller is usually designated in inches. A "78-52" propeller is 78 inches in length with an effective pitch of 52 inches. The effective pitch is the distance a propeller would move through the air in one revolution if there were no slippage. On a "78-52" propeller the distance would be 52 inches.

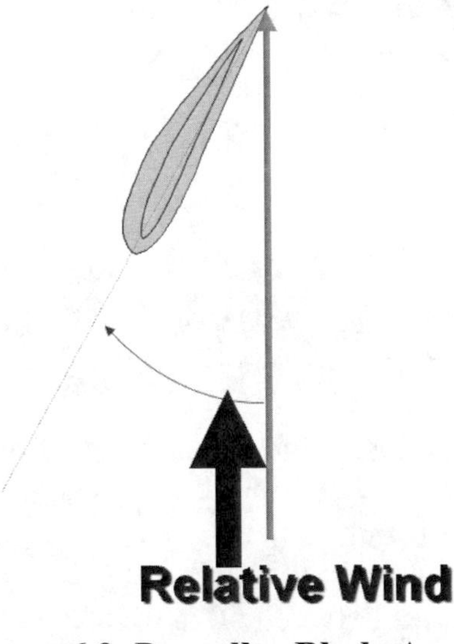

Figure 6.9. Propeller Blade Angle.
Source: Ben Trapnell, UND Aerospace

There are two types of propellers installed on most general aviation aircraft, a fixed pitch propeller and a controllable pitch propeller. The fixed pitch propeller is at one blade angle that will give it the best overall efficiency for the type of operation being conducted for which the aircraft was designed. For most aircraft this would be a cruise setting. A controllable pitch propeller allows the pilot to adjust the blade angle for the phase of flight. On takeoff and climb out a low pitch high RPM setting is used. During cruise flight a high pitch low RPM setting is generally used.

On the ground with the aircraft in a static condition the propeller efficiency is very low because each blade is moving through the air at an angle of attack which produces very low thrust to power ratio. This means that a lot of power is being used to sustain the engine and rotate the propeller and very little thrust is being produced. The propeller, unlike the wing, moves both rotationally and forward (dynamically). The angle at which the relative wind strikes the propeller blade is the AOA. This produces a higher dynamic pressure on the engine side which in turn is called thrust. Thus thrust is the relationship of propeller AOA and blade angle.

Since an aircraft moves forward through the air, it is important that the pilot understands how forward velocity affects the AOA of the propeller. Figure 6.10 shows the propeller in a static condition on the ground. At this point the relative wind is opposing propeller rotation. As forward velocity increases the relative wind moves closer to the chord line, increasing the propeller AOA—Figure 6.10. This can easily be demonstrated in an aircraft with a fixed pitch propeller by pitching up or down without changing power. When the aircraft is pitched down RPM will increase as the relative wind moves closer to the chord line and the AOA is decreased. When the aircraft is pitched up the RPM will decrease as the relative wind moves farther from the chord line and the AOA is increased.

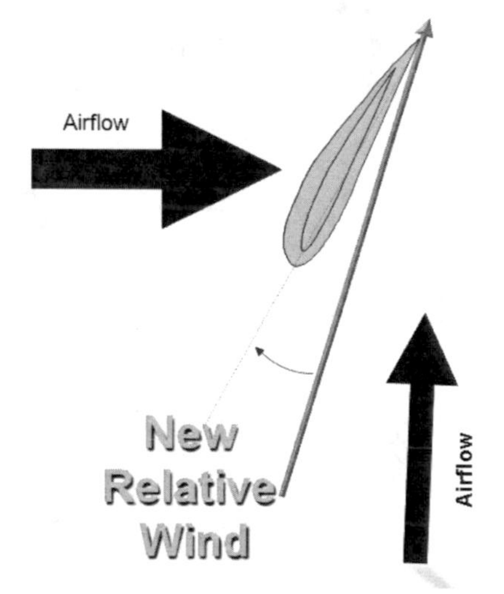

Figure 6.10. Propeller Blade Angle with Forward Velocity.
Source: Ben Trapnell, UND Aerospace

Propeller efficiency is a ratio between thrust horsepower and brake horsepower. Propeller efficiency usually varies between 50 and 80% on light general aviation aircraft. The measure of efficiency is how much a propeller slips in the air. This is measured by the geometric pitch (theoretical) which shows a propeller with no slippage. Effective pitch is the distance that the propeller actually travels—Figure 6.11.

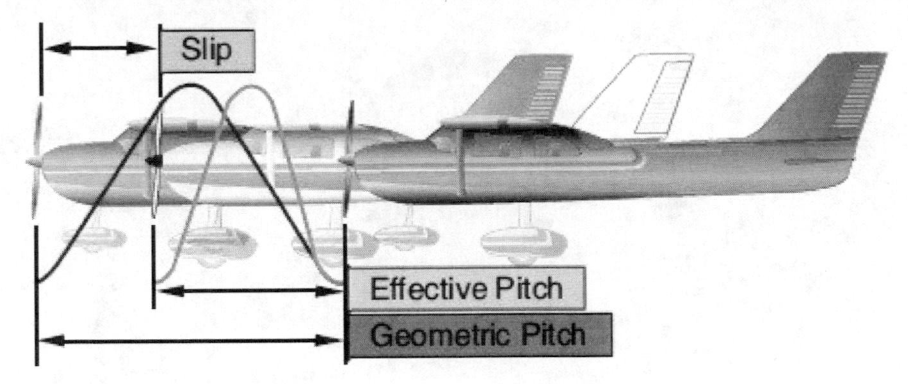

Figure 6.11. Propeller Efficiency.
Source: Pilot's Handbook of Aeronautical Knowledge, FAA

6.2.2 Propeller Twist

Another concept that warrants discussion is propeller twist. Because the tips of the propeller blades travel faster, due to their distance from the hub, the blades are twisted to even out the AOA along the propeller blade.

6.3 Left Turning Tendencies

The left turning tendencies on an aircraft have four elements which cause a form of rotation or twisting of the aircraft's axis. Left-turning tendencies can be split into four types:

1) Torque reaction
2) Asymmetric loading
3) Corkscrewing effect
4) Gyroscopic action

6.3.1 Torque Reaction

Torque reaction is explained by Newton's third law—for every action, there is an equal and opposite reaction. The propeller is rotating in one direction and an opposite, yet equal force is trying to rotate in the opposite direction—Figure 6.12.

Figure 6.12. Torque Reaction.
Source: Pilot's Handbook of Aeronautical Knowledge, FAA

In flight, this force is acting on the longitudinal axis making the aircraft roll. Trim or design features such as engine offset and wing twist can be used to counteract this rolling force.

On the ground torque is felt on the vertical axis. The left side of the aircraft is being forced down, creating more frictional force on the left tire when compared to the right. This can easily be corrected by using rudder and trim if available.

6.3.2 Asymmetric Loading

Asymmetric loading is the result of the descending blade being at a higher AOA than the ascending blade. This causes a yawing moment to the left—Figure 6.13.

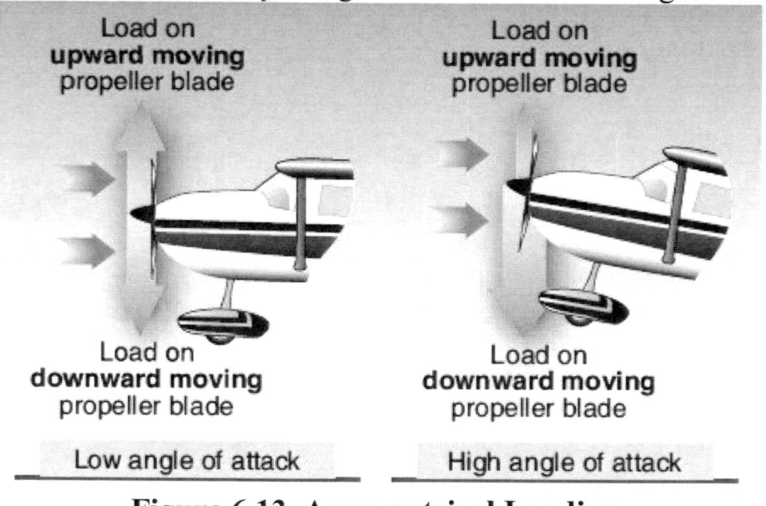

Figure 6.13. Asymmetrical Loading.
Source: Pilot's Handbook of Aeronautical Knowledge

6.3.3 Corkscrewing Effect

Corkscrewing effect is caused by the spiraling of the propeller's slipstream—Figure 6.14. This exerts a sideward force on the vertical tail surface and yaws the aircraft left about the vertical axis. The corkscrew slipstream also causes a rolling moment on the longitudinal axis. It should be noted that this rolling moment is to the right, which is opposite the rolling moment caused by torque.

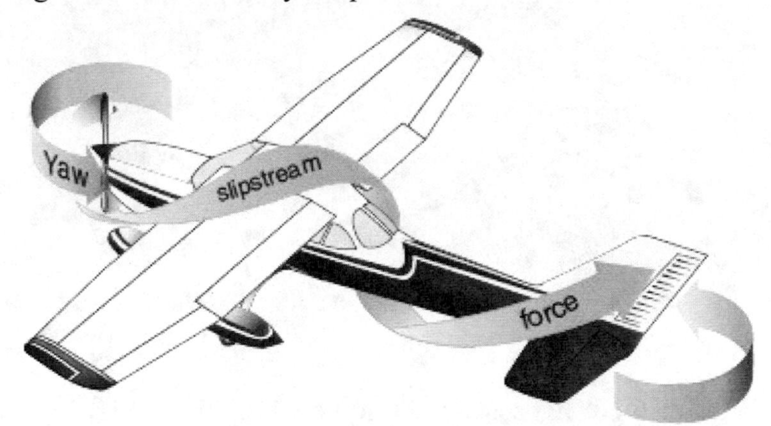

Figure 6.14. Corkscrewing Effect.
Source: Pilot's Handbook of Aeronautical Knowledge

6.3.4 Gyroscopic Action

Gyroscopic action is the result of a force being applied to a spinning disc. The resultant force is felt 90° ahead in the plane of rotation.

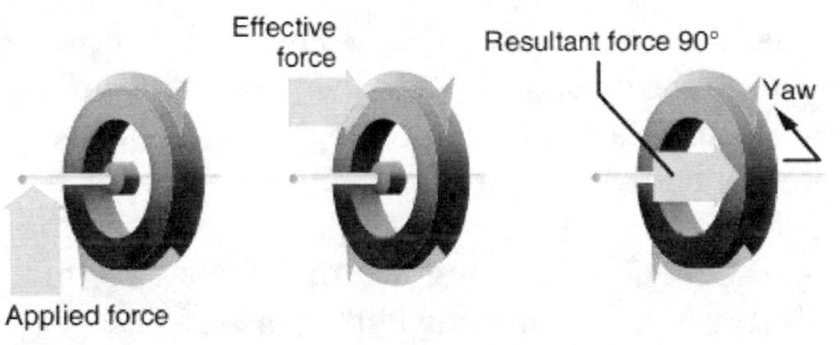

Figure 6.15. Precession.
Source: Pilot's Handbook of Aeronautical Knowledge

Precession is generally felt in tailwheel aircraft when the pilot raises the tail—Figure 6.16. The size of the force is determined by how fast the tail is raised. The yaw will occur on the vertical axis and the pilot will use the rudder to counteract the force.

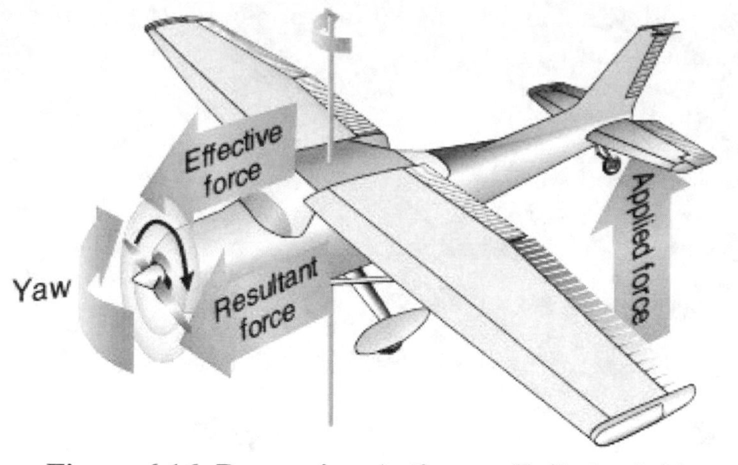

Figure 6.16. Precession Action on Tailwheel Aircraft.
Source: Pilot's Handbook of Aeronautical Knowledge

6.4 Climb Performance

Climb performance is a result of using the aircraft's potential energy provided by one or a combination of two factors. The first is the use of excess power above that required for level flight. An aircraft equipped with an engine capable of 200 horsepower (at a given altitude) but using 130 horsepower to sustain level flight (at a given airspeed) has 70 excess horsepower available for climbing. A second factor is that the aircraft can exchange its kinetic energy for potential energy by reducing its airspeed. The reduction in airspeed will increase the aircraft's potential energy, thereby also causing the aircraft to climb. Both terms, power and thrust, are often used in aircraft performance; however, they should not be confused. Although the terms "power" and "thrust" are sometimes used interchangeably, erroneously implying that they are synonymous, it is important to distinguish between the two when discussing climb performance.

Work is the product of a force moving an object a certain distance and is independent of time. Work is measured by several standards; the most common unit is a foot-pound. If a one pound mass is raised one foot, a work unit of one foot-pound has been performed. You can have force without work. For example, if you tried to push a 3000 lb airplane and exerted a force of 100 lbs against the aircraft and it did not move, force has been applied but no work has been accomplished. The common unit of mechanical power is horsepower; one horsepower (hp) is work equivalent to lifting 33,000 pounds a vertical distance of one foot in one minute. The term power implies work rate or units of work per unit of time, and as such is a function of the speed at which the force is applied. Thrust, also a function of work, means the force that imparts a change in the velocity of a mass. This force is measured in pounds but has no element of time or rate. It can then be said that, during a steady climb, *the angle of climb is a function of excess thrust.*

This relationship means that, for a given weight of an aircraft, the angle of climb depends on the difference between thrust and drag, or the excess power.

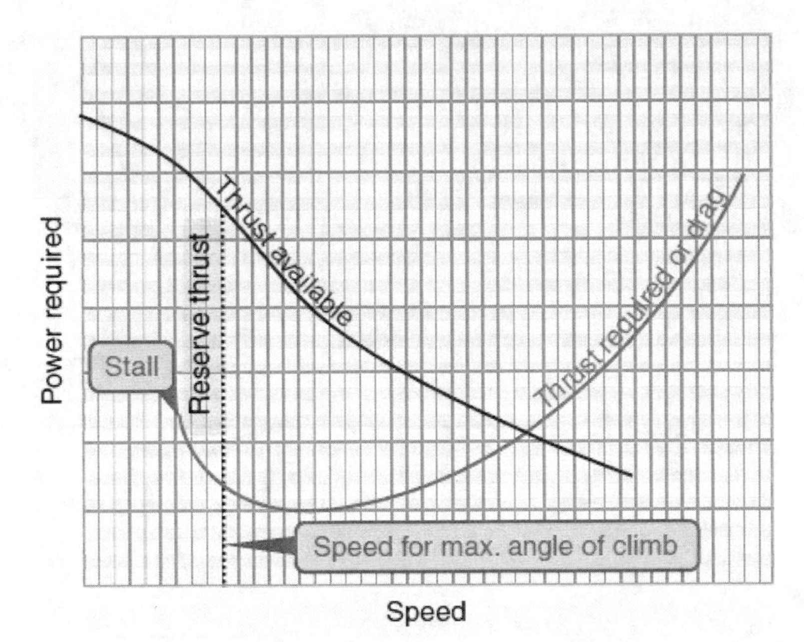

Figure: 6.17. Thrust Available versus Thrust Required.
Source: Pilot's Handbook of Aeronautical Knowledge

Of course, when the excess thrust is zero, the inclination of the flightpath is zero, and the aircraft will be in steady, level flight—Figure 6.17. When the thrust is greater than the drag, the excess thrust will result a climb angle depending on the value of the excess thrust. On the other hand, when the thrust is less than the drag, the deficiency of thrust will result in an angle of descent. The most immediate interest in the climb angle performance involves obstacle clearance. The most obvious purpose is to clear obstacles when climbing out of short or confined airports.

The forces acting on an airplane in climb are shown by Figure 6.18. The biggest difference can be seen with the aircraft's weight. The aircraft's weight acts perpendicular to the aircraft, but there is now a component that acts aft along the aircraft's flightpath. This rearward component of weight is added to the drag force of the aircraft and must be overcome by thrust.

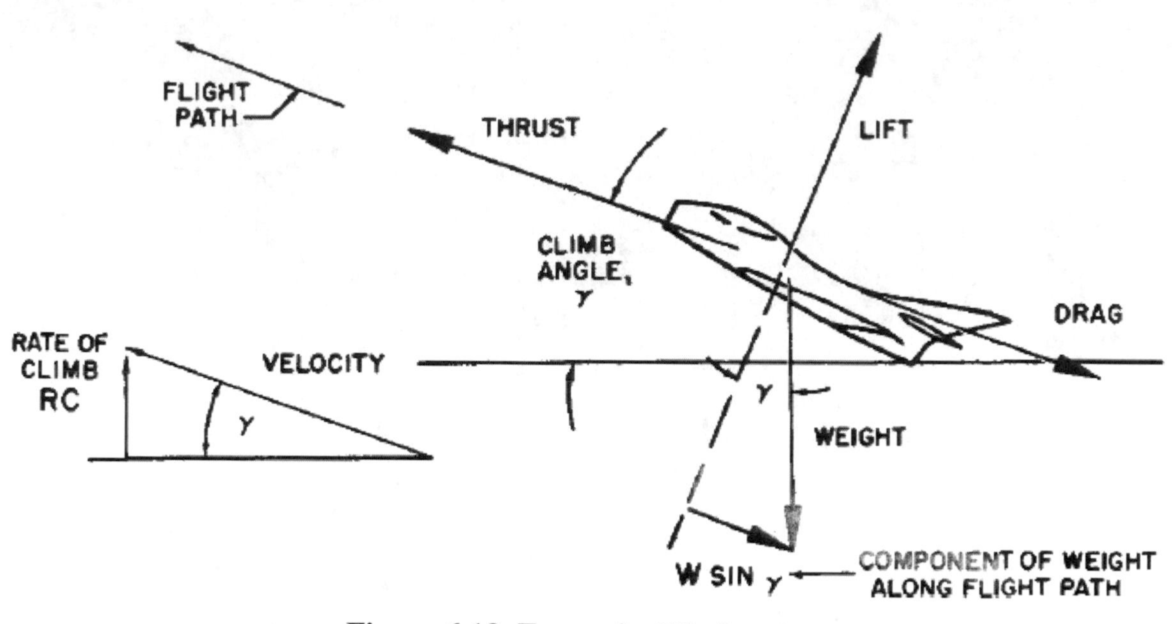

Figure 6.18. Forces in Climb.
Source: Aerodynamics for Naval Aviators

The second thing to discuss is the angle of attack versus the flightpath. Figure 6.19 shows the difference between the angle of attack and the flightpath. Notice that the thrust line is higher than the flightpath line. Because of this, the thrust which counteracts the weight must be greater than the weight due to this inclination. For example, if the aircrafts drag totaled 700 lb and your rearward component of weight totaled 200 lb; the total force acting rearward would be 900 lbs (drag and weight combined). The thrust required would have to be slightly higher to counteract the fact that the thrust vector is above the flightpath.

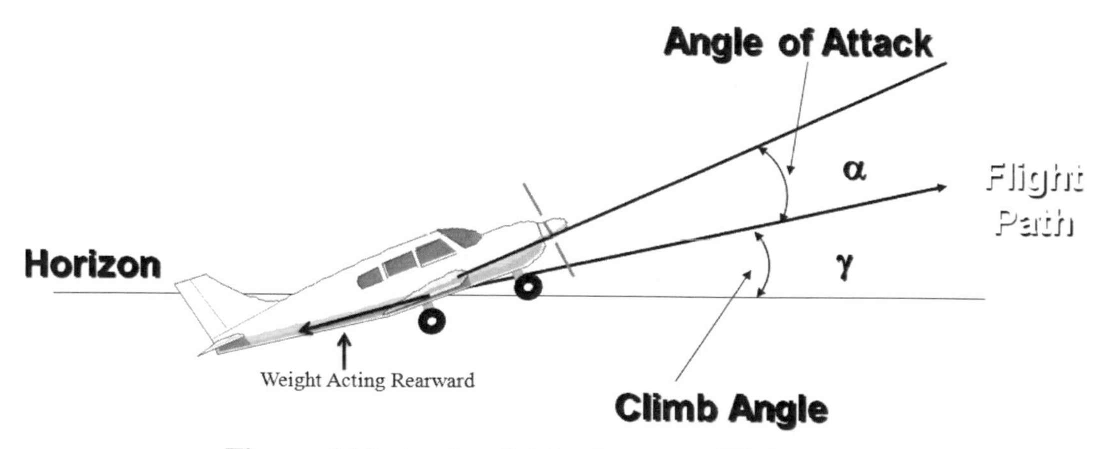

Figure 6.19. Angle of Attack versus Flightpath.
Source: Tom Zeidlik, UND Aerospace

Another portion of thrust is actually aiding lift; it is called the vertical component of thrust. Figure 6.20 shows that the vertical component of thrust is acting 90° to the thrust line, or parallel to lift.

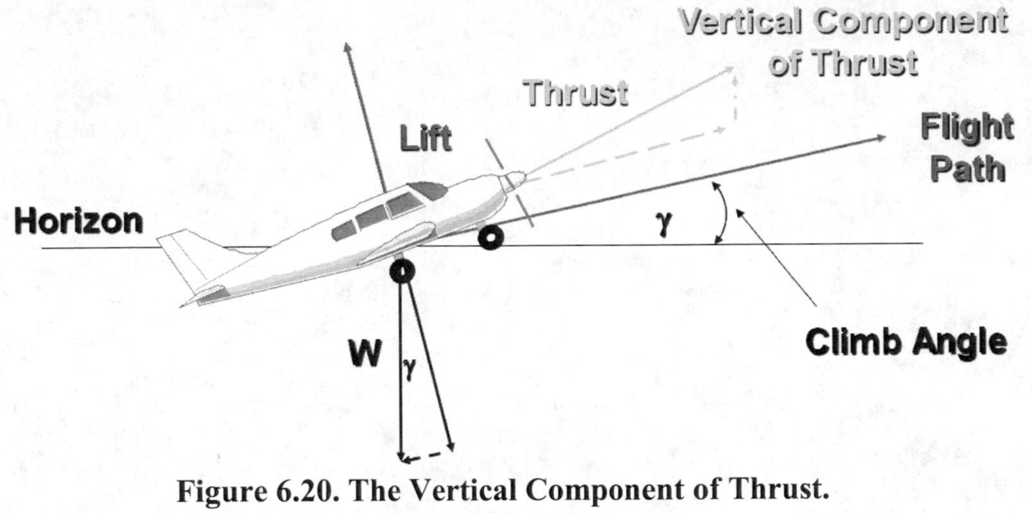

Figure 6.20. The Vertical Component of Thrust.
Source: Tom Zeidlik, UND Aerospace

6.4.1 Aircraft Climb Rate

One of the most basic things a pilot wants to know is what the rate of climb will be in an aircraft. This can be determined using the following equation:
RC = Rate of climb
Pa = Power available
Pr = Power required
W – Weight

$$RC = Pa - \text{Pr}\, x \frac{33{,}000}{W}$$

Pa and Pr are determined using a power available versus a power required chart—Figure 6.21. From a practical point of view pilots do not calculate rate of climb using this method, they will refer to their respective Pilot's Operating Handbook section 5 and use the rate of climb chart. This requires data (weight, temperature, and pressure altitude) that are readily available to the pilot.

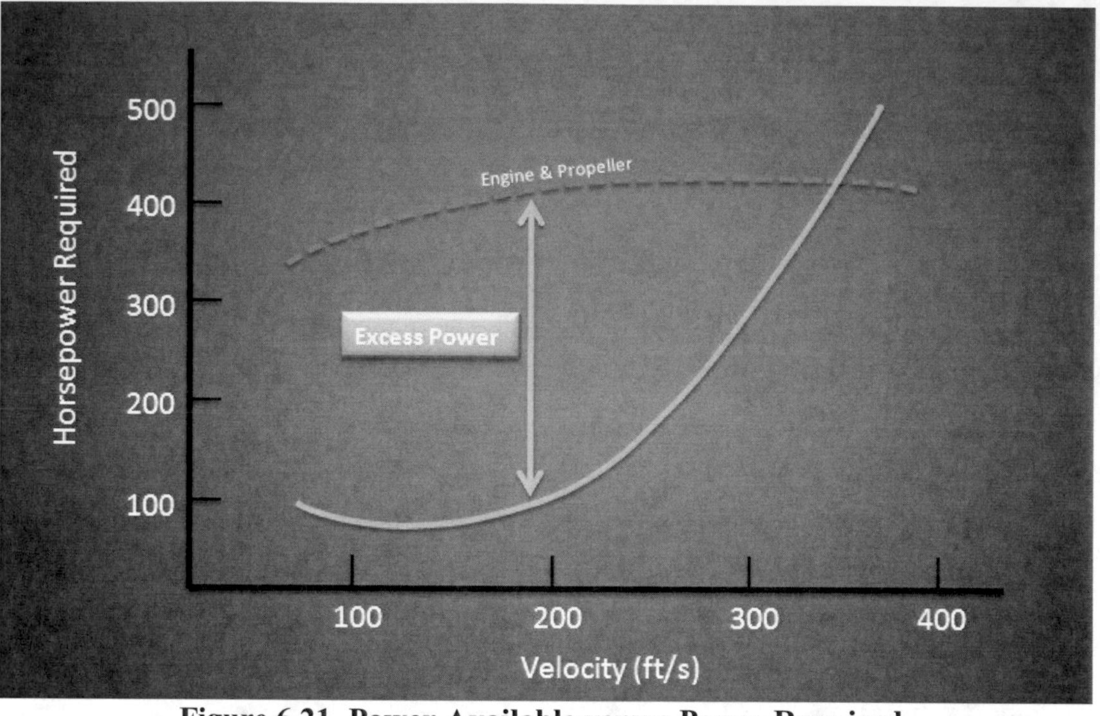

Figure 6.21. Power Available versus Power Required.
Source: Shelby Balogh, UND Aerospace

The maximum angle of climb is the point where there is the greatest difference between thrust available and thrust required; for example, for the propeller-powered airplane, the maximum excess thrust and angle of climb will occur at some speed just above the stall speed. Thus, if it is necessary to clear an obstacle after takeoff, the propeller-powered airplane will attain maximum angle of climb at an airspeed close to—if not at—the takeoff speed.

Of greater interest in climb performance are the factors that affect the rate of climb. The vertical velocity of an aircraft depends on the flight speed and the inclination of the flightpath. In fact, the rate of climb is the vertical component of the flightpath velocity.

The maximum rate of climb would occur where there exists the greatest difference between power available and power required—Figure 6.22.

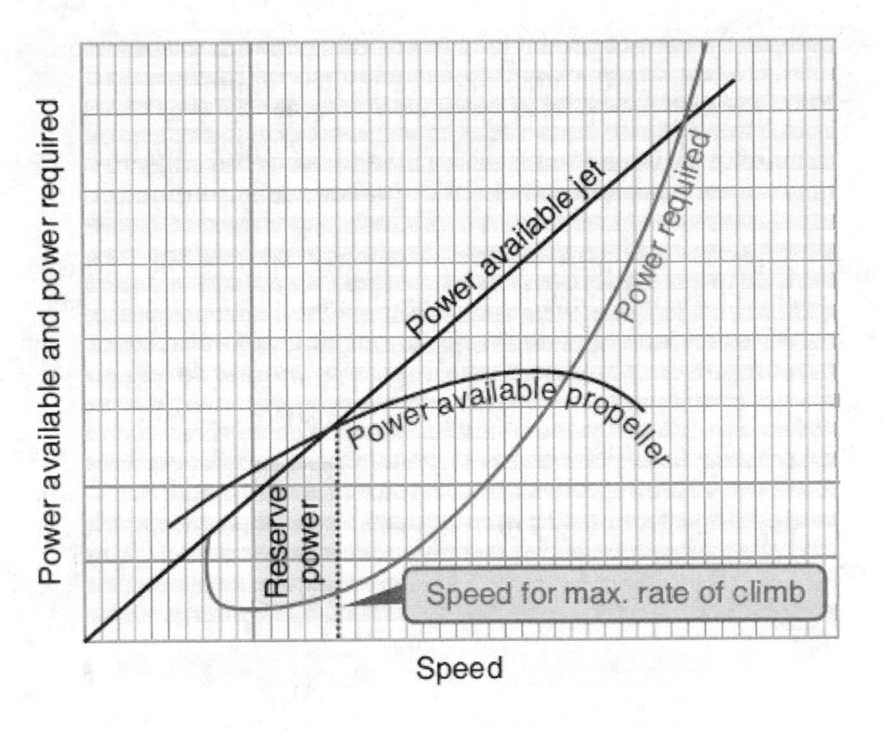

Figure 6.22. Maximum Rate of Climb.
Source: Pilot's Handbook of Aeronautical Knowledge

The above relationship means that, for a given weight of an aircraft, the rate of climb depends on the difference between the power available and the power required, or the excess power. Of course, when the excess power is zero, the rate of climb is zero and the aircraft is in steady, level flight. When power available is greater than the power required, the excess power will allow a rate of climb specific to the magnitude of excess power. During a steady climb, the rate of climb will depend on excess power while the angle of climb is a function of excess thrust.

The climb performance of an aircraft is affected by certain variables. The conditions of the aircraft's maximum climb angle or maximum climb rate occur at specific speeds, and variations in speed will produce variations in climb performance. There is sufficient latitude in most aircraft that small variations in speed from the optimum do not produce large changes in climb performance, and certain operational considerations may require speeds slightly different from the optimum. Of course, climb performance is most critical at high gross weight, at high altitude, in obstructed takeoff areas, or during malfunction of a powerplant. In these conditions maintaining optimum climb speed is necessary.

Weight has a very pronounced effect on aircraft performance. If weight is added to an aircraft, it must fly at a higher angle of attack (AOA) to maintain a given altitude and speed. This increases the induced drag of the wings, as well as the parasite drag of the aircraft. Increased drag means that additional thrust is required which results in less excess thrust being available for climbing. Aircraft designers go to great effort to minimize the aircraft weight since it has such a marked effect on the factors pertaining to performance.

A change in an aircraft's weight produces a twofold effect on climb performance. First, a change in weight will change the drag and the power required. This alters the excess power available, which in turn, affects both the climb angle and the climb rate. Secondly, an increase in weight will reduce the maximum rate of climb, but the aircraft must be operated at a higher climb speed to achieve the smaller peak climb rate.

An increase in altitude also will increase the power required and decrease the excess power available. Therefore, the climb performance of an aircraft diminishes with altitude. The speeds for maximum rate of climb, maximum angle of climb, and maximum and minimum level flight airspeeds vary with altitude. As altitude is increased, these speeds converge at the **absolute ceiling** of the aircraft. At the absolute ceiling, there is no excess power available, and only one speed will allow steady, level flight. Consequently, the absolute ceiling of an aircraft produces zero rate of climb. The **service ceiling** is the altitude at which the aircraft is unable to climb at a rate greater than 100 feet per minute (fpm). Usually, these specific performance reference points are provided for the aircraft at a specific design configuration—Figure 6.23

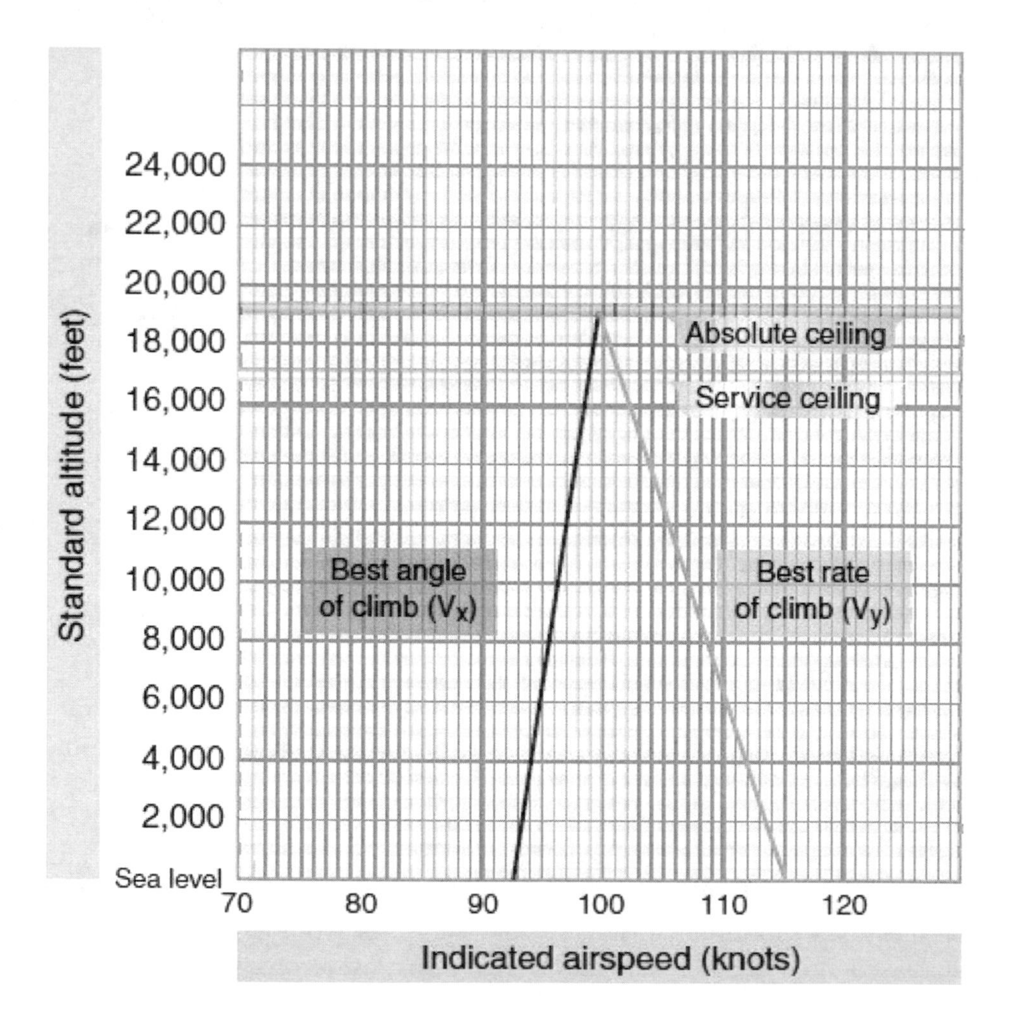

Figure 6.23. Performance Reference Points.
Source: Pilot's Handbook of Aeronautical Knowledge

The absolute ceiling of the aircraft serves no practical purpose; a pilot could only obtain it at one critical airspeed where the horsepower available and the horsepower required intersect at one point. On the other hand the service ceiling is attainable, but also not a practical altitude in the national airspace system. If a pilot chooses cruise altitudes close to the service ceiling of the aircraft he or she will spend a large portion of the flight climbing and will reduce efficiency and add more time to the leg.

6.4.2 Best Angle of Climb

V_x is the **best angle of climb airspeed**. In simple terms, it is the greatest gain in altitude over a horizontal distance. This is the speed a pilot would fly at to clear obstacles after takeoff or during flight. Looking at this speed in the aircraft's flight manual we see that short field charts do not actually publish the V_x speed on the chart. This is because as the aircraft accelerates to rotation speed and then lifts off—it takes time to accelerate to V_x. This is why a 50 ft speed is usually published and is used as a reference as the pilot accelerates to V_x.

V_x changes with altitude because the angle of climb is a function of excess thrust, which changes with both altitude and weight. As indicated in the chart above (Figure 6.23), we see V_x increases with altitude. If we plotted a rate of climb versus true airspeed graph we would see that as we change altitude or aircraft weight the rate of climb chart would change.

6.4.3 Best Rate of Climb

V_y is the **best rate of climb airspeed** and is a function of excess power. V_y in simple terms is the greatest gain in altitude over time. V_y is the speed that is used predominately in most light general aviation aircraft during climb. V_y as an indicated airspeed decreases with altitude. As a true airspeed it will actually increase. This is an important point because most graphs that show power required versus power available have the speed along the x-axis labeled as true airspeed. This creates confusion because the discussion centers on V_y decreasing with altitude, but most chart show it increasing with altitude.

V_y decreases with altitude because the power required curve and power available curve change with altitude and weight.

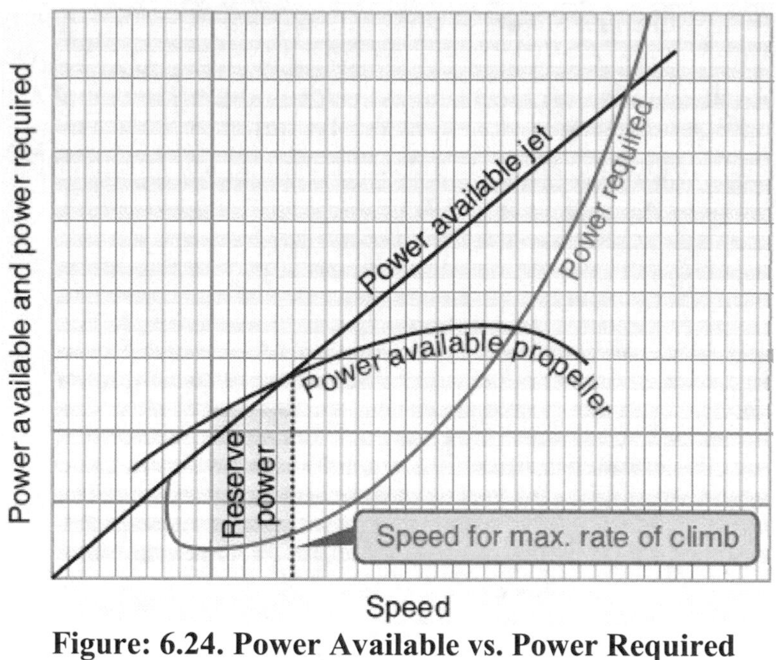

Figure: 6.24. Power Available vs. Power Required
Source: Pilot's Handbook of Aeronautical Knowledge

As the aircraft climbs in altitude the air becomes less dense, this causes the power required curve to shift up and rotate to the right—Figure 6.25.

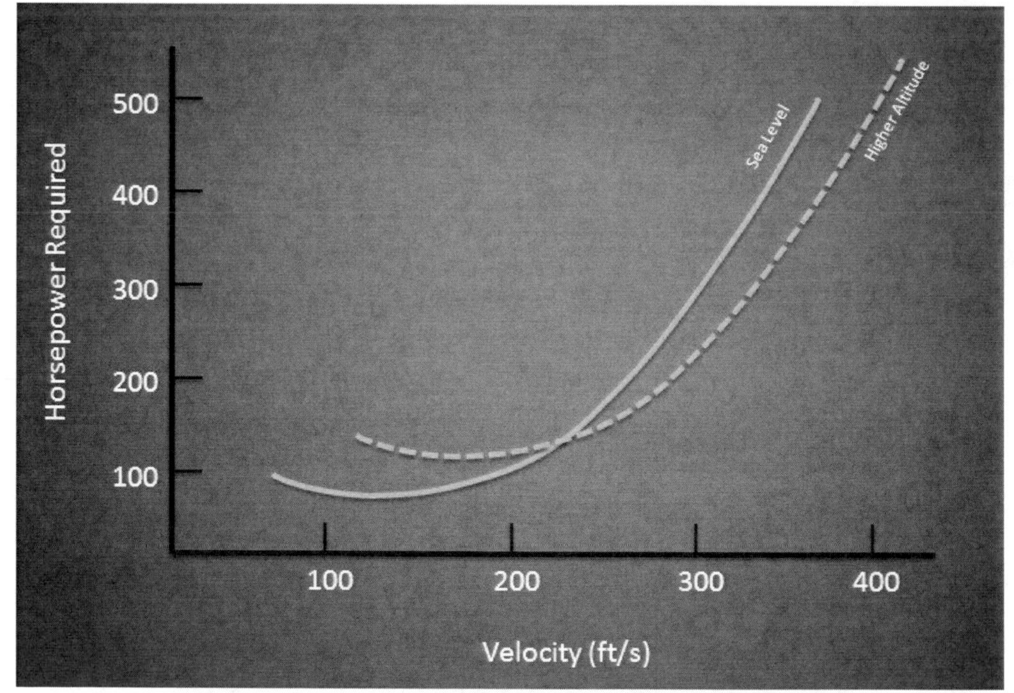

Figure 6.25. Velocity vs. Horsepower Required
Source: Shelby Balogh, UND Aerospace

Remember the power required curve is nothing more than the total drag curve plotted in terms of the minimum power required. Induced drag will increase at all speeds as the aircraft goes up in altitude because the air is less dense; this requires the pilot to increase the angle of attack to produce enough lift to counteract weight. Parasite drag on the other hand will decrease across all speeds as the air becomes less dense because there is less frictional force.

6.5 Cruise Flight

Cruise flight centers on two basic principles, how far we can fly, and for how long. How far we can fly is defined as the aircraft's **range**. How long we can fly is defined as **endurance**.
When we go flying we generally consider range in two ways:

1) Maximizing the distance we fly for a given fuel load.
2) Traveling a specified distance while burning minimum fuel

6.5.1 Endurance

This discussion should start by making sure you understand that range and endurance are not the same. Range relates to distance, endurance relates to time. The formula for endurance is:

$$endurance = \frac{hours}{fuel}$$

Hours is simply flight time expressed in whatever units you want. Fuel can be expressed in gallons or pounds. If a pilot wants to achieve maximum endurance he or she would slow the aircraft to the minimum powered required speed. Figure 6.26 shows the minimum power point being the lowest point in the drag curve.

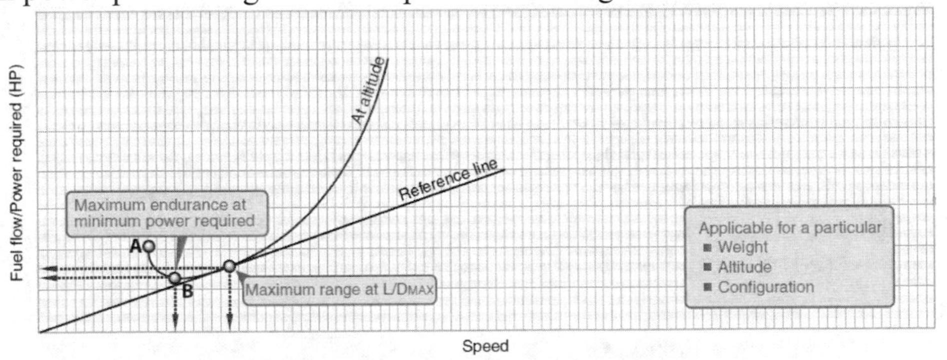

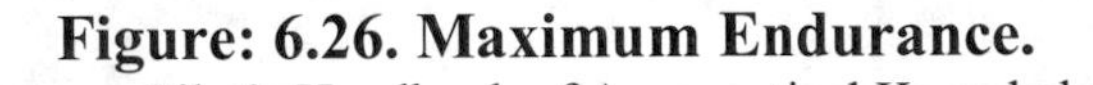
Figure: 6.26. Maximum Endurance.
Source: Pilot's Handbook of Aeronautical Knowledge

If the aircraft were to slow even further, to point A, drag would increase rapidly, more power would be required, and the engine would burn more fuel. If the aircraft were to accelerate above point B, drag also increases, which increases fuel burn. As you can see, flying at maximum endurance speed in not practical in the real world—you may save fuel but it would take forever to get to the destination. This speed is also not practical for operations such as holding because it is generally close to stall. Endurance from a practical standpoint comes from the selection of a cruise power setting of 55%, 65%, or 75% endurance charts. The point of this type of flying is generally to minimize or eliminate fuel stops (very time consuming) along the route, or to minimize fuel burn for cost purposes—not necessarily to stay aloft for hours on end.

6.5.2 Range

Range can be broken down into two parts, specific and total range. An easy way to understand this example is to use a car trip scenario. If I have a car that has a 20 gallon fuel tank and I get 30 miles per gallon I can travel 600 miles on one tank of gas. The specific range in this example is 30 miles per gallon, the total range is 600 miles. In an airplane **specific range** would be how many nautical miles you can travel on one gallon or pound of fuel. The **total range** would be how far the airplane can fly with the remaining fuel load on board the aircraft. The definition for specific range is:

$$specific\ range = \frac{nm}{gallons\ of\ fuel}$$

Note: pounds can be inserted for gallons

Specific range is affected by three things: 1) aircraft weight, 2) altitude, and 3) configuration. The **maximum range** of the aircraft can be found at L/D_{max}. Unlike endurance, which is found on the drag curve where minimum power is required, maximum range is found where the ratio of speed to power required is the greatest. This is located on the graph by drawing a tangent line from the origin to the power required curve—Figure 6.26. Another way to think about this is that as you move from the origin point along the tangent line toward L/D_{max} you increase airspeed at a greater rate than fuel burn (think of the ratio). At L/D_{max} the ratio of fuel to airspeed should be 1. At any speed above L/D_{max}, the fuel burn ratio increases at a greater rate than the airspeed. Therefore, L/D_{max} is the point where the speed to power ratio required is the greatest.

Another aspect of range that we need to look at is the effect of weight on range. Because L/D_{max} occurs at a specific angle of attack, and most general aviation airplanes do not have AOA indicators, the airspeed has to be varied as weight changes to maintain a constant AOA. Figure 6.27 shows this: as weight increases, the speed must be increased to maintain the AOA. This is because as weight is increased the AOA must be increased to produce more lift; the only way to lower the AOA is to increase speed. As weight decreases, the speed must decrease. The reasoning is that, as the aircraft becomes lighter, the AOA is lowered to compensate for less weight; the only way to increase AOA is to reduce speed—Figure 6.27.

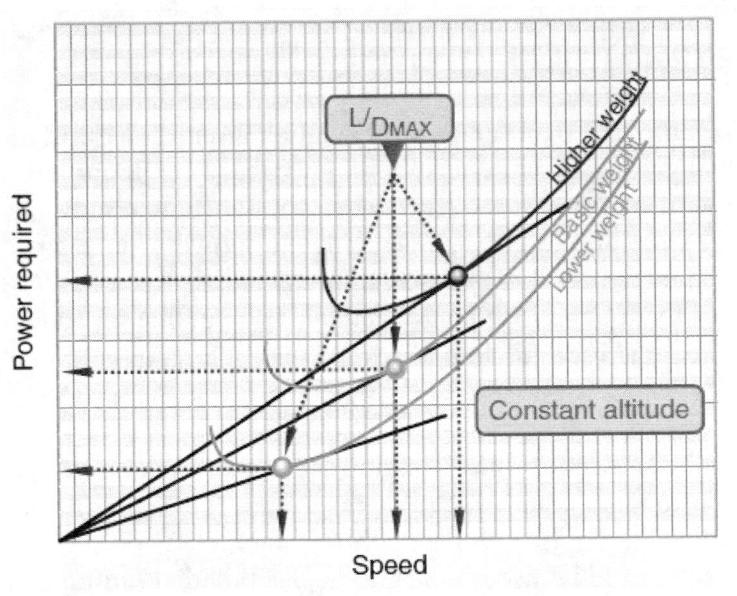

Figure: 6.27. Weight Effects on Range.
Source: Pilot's Handbook of Aeronautical Knowledge

The effect of altitude on range can be seen by referring to Figure 6.28. Flights operating at high altitude require a higher TAS, which will require more power.

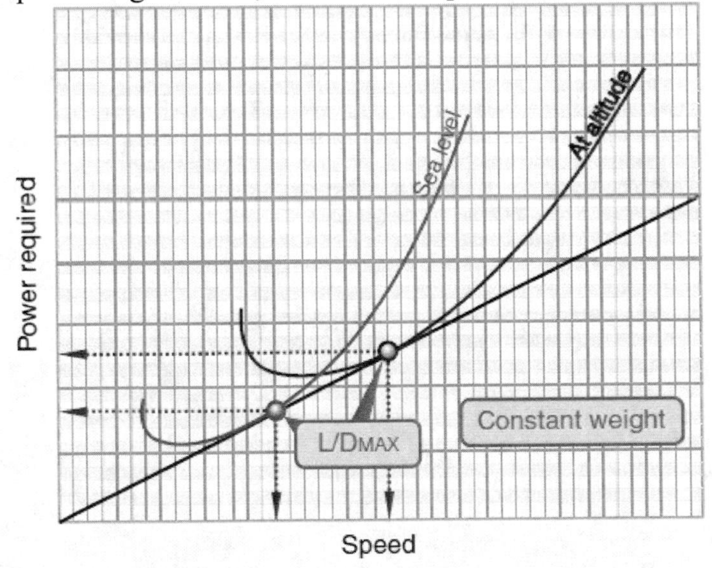

Figure 6.28. Effect of Altitude on Range.
Source: Pilot's Handbook of Aeronautical Knowledge

Another aspect of cruise flight that is often not talked about in textbooks regarding range and endurance is cruise performance. From a practical standpoint the pilot will not fly the aircraft at maximum endurance or range—it is just too slow. In reality is we often operate the propeller-driven airplane at 55%, 65%, or 75% best power or endurance. In order to calculate how to get to your destination as fast as possible, find the highest true airspeed for your aircraft. Most fixed gear single engine aircraft that cruise in the 110–130 knot range will have their highest TAS in the 6000 to 7000 ft range. This is a good place to start; however, the wind, terrain, and the need for fuel stops will dictate the altitude and speed at which the aircraft ultimately fly.

6.6 Descent and Gliding Flight

Descending a light propeller-driven general aviation aircraft is a fairly simple task. Reduce power to a point where there is more power required than power available and the basic principle of weight takes over. Under normal flight conditions descending flight is initiated by the pilot creating a decrement of power (more power required than available). Once the aircraft begins descending the weight vector can be broken up into two parts, just like the climb. One component acts perpendicular (down) to the flightpath, the other acts forward and parallel to the flightpath, helping accelerate the aircraft.

Gliding flight can be self induced by bringing the power back to idle, but in most piston aircraft, descents are not conducted at idle power, thus they are called a powered descent. This is because of shock cooling and the possible damage it could cause to the engine. A true gliding descent would be used if the engine has failed. Gliding flight can be broken down into two parts, minimum sink or maximum range.

The **minimum sink glide** is used to prolong your time aloft in the event the engine or engines fail. This is a speed that is not published, but could be useful if you are over your current landing site and wish to stay aloft a little longer. Most light single engine airplanes will be at minimum sink (or close to) with full aft trim. This is slower than best glide speed. If a pilot elects to use this method he or she should accelerate to best glide once they reach a normal pattern altitude. This will provide a larger margin above stall and the aircraft will have more positive maneuverability. It should be noted that the best glide speed should be used unless the pilot has training and experience flying at the minimum sink glide speed.

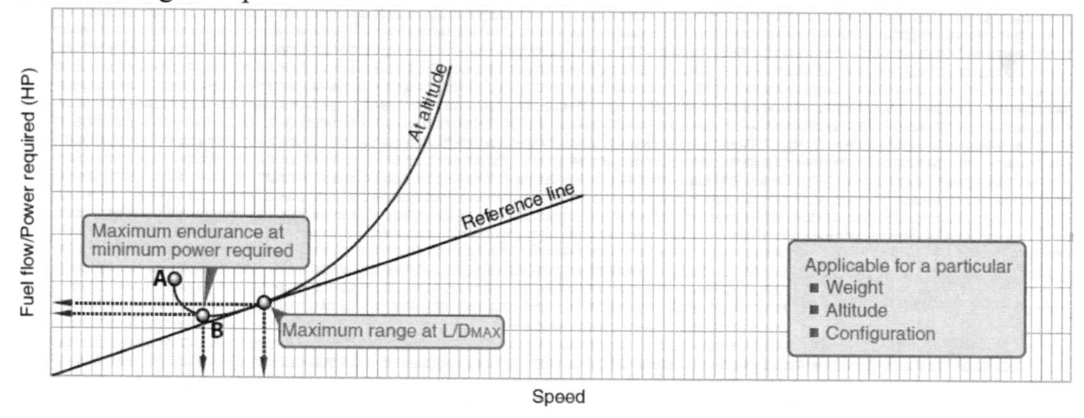

Figure 6.29. Maximum Range & Endurance Chart
Source: Pilot's Handbook of Aeronautical Knowledge

The **maximum glide range** occurs at the speed for maximum range – L/D_{max}. This is generally a published speed and is used when the engine stops or fails in flight. Some Airplane Flying Manuals (AFMs) contain glide ratio charts. There are some concerns with these charts:

1) They do not account for wind.
2) They are usually calculated in a minimum drag configuration (gear and flaps up).

3) They are usually calculated with controllable pitch propellers in the full decrease position (high AOA).
4) They are usually calculated at maximum gross weight.

Wind is a factor in glide distance and angle. A headwind will decrease glide distance, and the angle of descent will increase (steepen). A tailwind will increase the glide distance and flatten the angle of descent. You experience the effects of both a headwind and a tailwind when you do a power off approach. On downwind the aircraft has a flatter descent and a higher groundspeed. Turning base to final the angle of descent steepens and the groundspeed slows.

Weight is also a factor in glide distance if L/D_{max} is not maintained. Without an AOA indicator the only way to maintain a specific AOA at L/D_{max} is to vary the airspeed. As weight increases, the airspeed would need to be increased to maintain L/D_{max}.

Altitude also affects the airplane's gliding distance. To understand this we need to step back and look at the effects of altitude on true airspeed. As the aircraft climbs, TAS increases about 2% per 1000 feet. An aircraft gliding at higher altitudes will have a higher TAS, this means that it will be moving down the slope at a faster rate. This is of particular importance when operating an aircraft at high density altitudes.

6.7 Takeoff and Landing

The majority of pilot-induced aircraft accidents occur during the takeoff and landing phases of flight. The pilot must be familiar with all the variables that influence the takeoff and landing performance of an aircraft and must strive for exacting, professional procedures of operation during these phases of flight.

Takeoff and landing performance is a condition of accelerated and decelerated motion. For instance, during takeoff, an aircraft starts at zero speed and accelerates to the takeoff speed to become airborne. During landing, the aircraft touches down at the landing speed and decelerates to zero speed. The important factors of takeoff or landing performance are:

- The takeoff or landing speed is generally a function of the stall speed or minimum flying speed.
- The rate of acceleration/deceleration during the takeoff or landing roll. The speed (acceleration and deceleration) experienced by any object varies directly with the imbalance of forces and inversely with the mass of the object. An airplane on the runway moving at 75 knots has four times the energy it has traveling at 37 knots. Thus, an airplane requires four times as much distance to stop as required at half the speed.
- The takeoff or landing roll distance is a function of both acceleration/deceleration and speed.

6.7.1 Runway Surface and Gradient

Runway conditions affect takeoff and landing performance. Typically, performance charts assume paved, level, smooth, and dry runway surfaces. Since no two runways are identical, the runway surface differs from one runway to another, as does the runway gradient or slope.

Runway composition also varies widely from one airport to another. The runway surface encountered may be concrete, asphalt, gravel, dirt, or grass. The runway surface for a specific airport is noted in the Airport/Facility Directory (A/FD). Any surface that is not hard and smooth will increase the ground roll required for takeoff. This is due to the added frictional force caused by the interaction of the tires and the runway surface. Tires can sink into soft, grassy, or muddy runways. Potholes or other ruts in the pavement can be the cause of poor tire movement along the runway. Obstructions such as mud, snow, or standing water reduce the airplane's acceleration down the runway. Although muddy and wet surface conditions can reduce friction between the runway and the tires, they can also act as obstructions and reduce the landing distance. Braking effectiveness is another consideration when dealing with various runway types. The condition of the surface affects the braking ability of the airplane.

The amount of power that is applied to the brakes without skidding the tires is referred to as **braking effectiveness**. Ensure that runways are adequate in length for takeoff acceleration and landing deceleration when less than ideal surface conditions are being reported.

The **gradient or slope of the runway** is the amount of change in runway height over the length of the runway. The gradient is expressed as a percentage such as a 3 percent gradient. This means that for every 100 feet of runway length, the runway height changes by 3 feet. A positive gradient indicates the runway height increases, and a negative gradient indicates the runway decreases in height. An upsloping runway impedes acceleration and results in a longer ground run required for takeoff. However, landing on an upsloping runway typically reduces the landing roll. A downsloping runway aids in acceleration on takeoff resulting in shorter takeoff distances. The opposite is true when landing, as landing on a downsloping runway increases landing distances. Runway slope information is contained in the A/FD.

6.7.2 Water on the Runway and Dynamic Hydroplaning

Water on the runways reduces the friction between the tires and the runway surface, and can reduce braking effectiveness. The ability to brake can be completely lost when the tires are hydroplaning because a layer of water separates the tires from the runway surface. This is also true when runways are covered in ice.

When the runway is wet, the pilot may be confronted with dynamic hydroplaning. **Dynamic hydroplaning** is a condition in which the aircraft tires ride on a thin layer of water rather than on the runway's surface. Because hydroplaning wheels are not touching the runway, braking and directional control are almost zero. To help minimize dynamic hydroplaning, some runways are grooved to help drain off water.

Tire pressure is a factor in dynamic hydroplaning. Using the simple formula $9\sqrt{TirePressure}$ each pilot can calculate the minimum speed, in knots, at which hydroplaning will begin. In plain language, the minimum hydroplaning speed is determined by multiplying the square root of the main gear tire pressure in psi by nine. For example, if the main gear tire pressure is at 36 psi, the aircraft would begin hydroplaning at 54 knots.

Landing at higher than recommended touchdown speeds will expose the aircraft to a greater potential for hydroplaning. Once hydroplaning starts, it can continue well below the minimum initial hydroplaning speed.

On wet runways, directional control can be improved by landing into the wind. Abrupt control inputs should be avoided. When the runway is wet, anticipate braking problems well before landing and be prepared for hydroplaning. Opt for a suitable runway most aligned with the wind. Mechanical braking may be ineffective, so aerodynamic braking should be used to its fullest advantage.

6.7.3 Takeoff Performance

The minimum takeoff distance is of primary interest in the operation of any aircraft because it defines the runway length requirements. The minimum takeoff distance is obtained by taking off at some minimum safe speed that allows sufficient margin above stall and provides satisfactory control and initial rate of climb. Generally, the lift-off speed is a fixed percentage of the stall speed or minimum control speed for the aircraft in the takeoff configuration. As such, the lift-off will be accomplished at a particular value of lift coefficient and AOA. Depending on the aircraft characteristics, the lift-off speed will be anywhere from 1.05 to 1.25 times the stall speed or minimum control speed.

To obtain minimum takeoff distance at the specific lift-off speed, the forces that act on the aircraft must provide the maximum acceleration during the takeoff roll. Newton's second law defines acceleration:

$$A = \sum F/M$$

The various forces acting on the aircraft may or may not be under the control of the pilot, and various procedures may be necessary in certain aircraft to maintain takeoff acceleration at the highest value. The forces acting on the aircraft are illustrated in Figure 6.30.

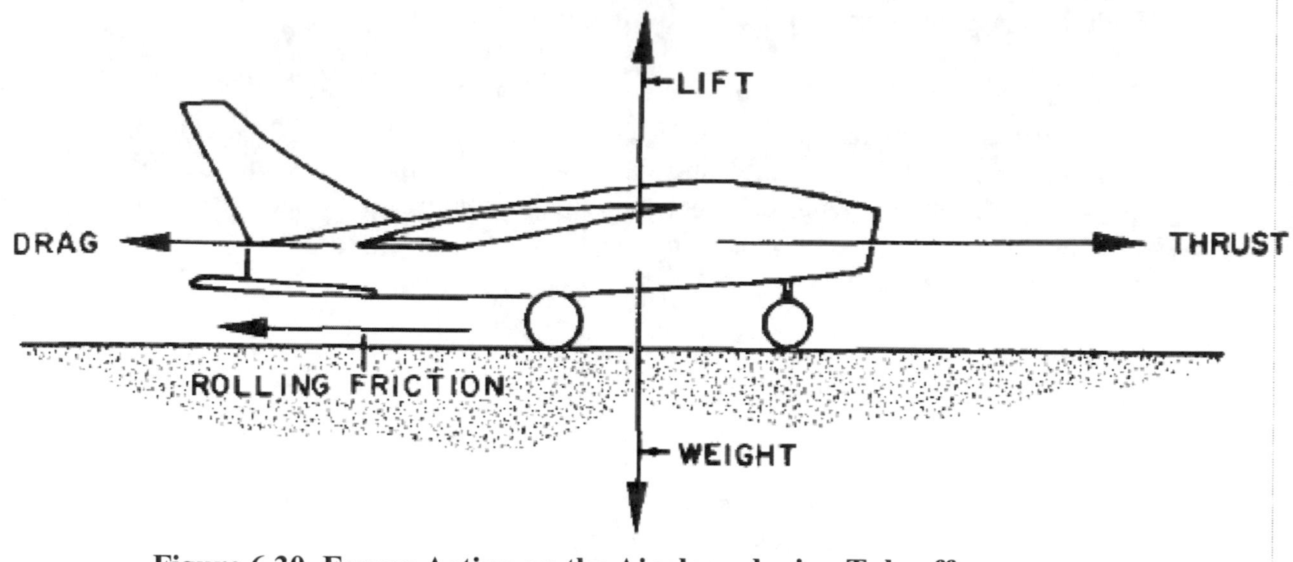

Figure 6.30. Forces Acting on the Airplane during Takeoff.
Source: Aerodynamics for Naval Aviators

The forces during takeoff can be broken down into three parts: 1) thrust, 2) drag, and 3) friction. Lift and weight are also important because they affect rolling resistance.

Thrust is the forward acting force and is key to accelerating the aircraft. The more thrust the propeller or engine produces the faster the aircraft will accelerate. For propeller-driven aircraft thrust decreases as velocity increases—Figure 6.31.

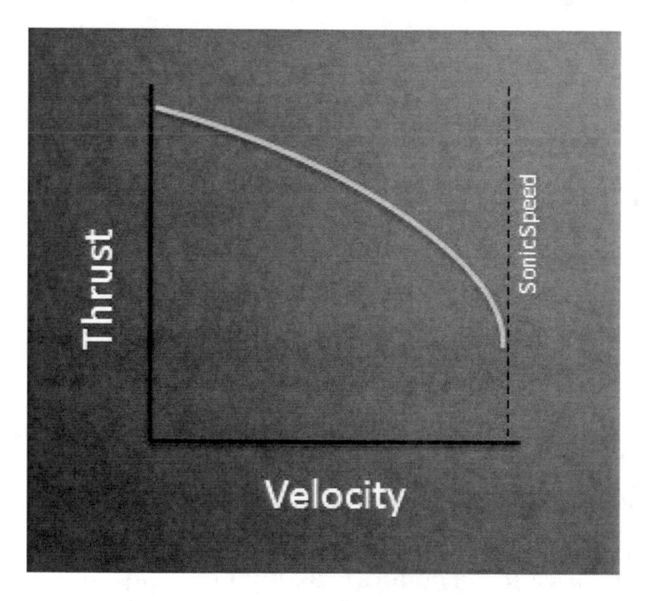

Figure 6.31. Thrust versus Velocity – Propeller-driven Airplane.
Source: Shelby Balogh, UND Aerospace

Drag, along with rolling friction, resists the acceleration of the aircraft. Drag acts rearward, rolling friction acts downward. Drag is produced as soon as the aircraft is moving. Aerodynamic drag increases as velocity increases (Figure 6.32). Drag can be broken down into two types, parasite and induced. Parasite drag predominates at low velocities; both act equally at high velocities. Rolling friction results when there is a normal force on the wheels and the frictional force is the product of the normal force (W-L) and the coefficient of rolling friction. The coefficient of friction is different for different surfaces. Rolling resistance decreases as velocity increases:

Surface	Coefficient
Dry Concrete	0.7
Light Rain	0.5
Heavy Rain	0.3
Snow or Ice	0.1-0.2

Average rolling coefficient of a tire is 0.2

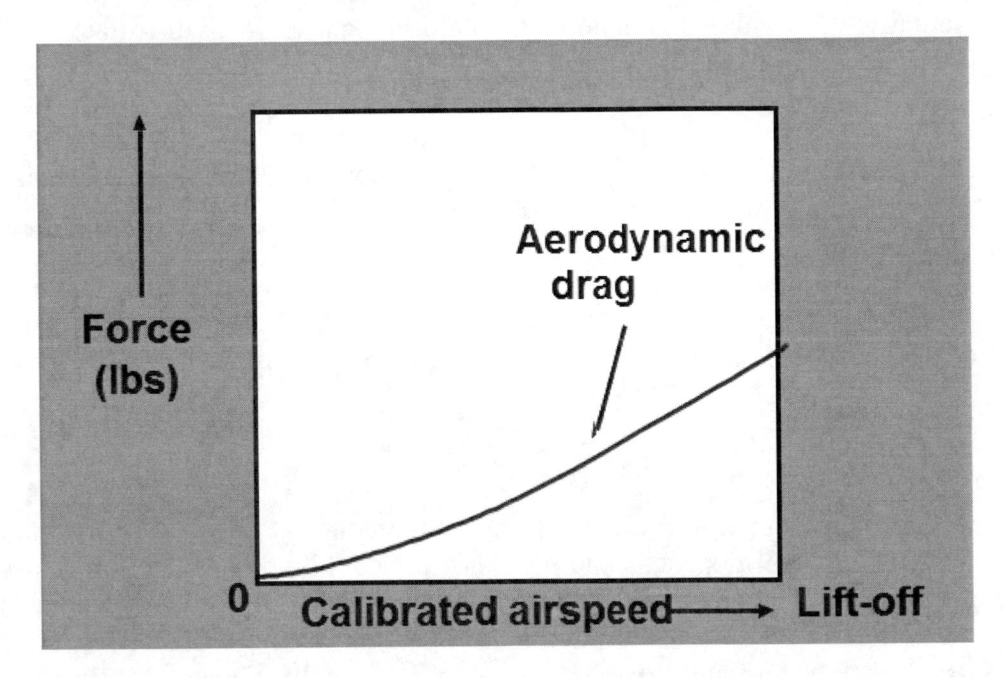

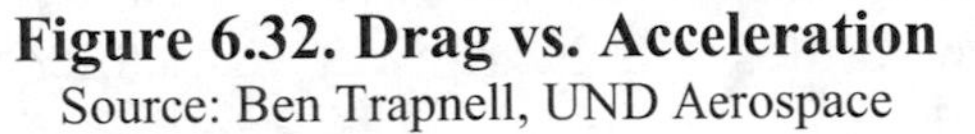
Figure 6.32. Drag vs. Acceleration
Source: Ben Trapnell, UND Aerospace

Acceleration during the takeoff roll can be determined by using the following equation:

$$a = \frac{g\{T - D - \mu(W - L)\}}{W}$$

To minimize takeoff distance we need to maximize thrust. This can be accomplished by increasing thrust, reducing drag, reducing weight, or increasing lift. To minimize ground roll, weight must be decreased and the coefficient of lift, thrust, and the surface area of the wing must be maximized.

6.7.4 Variables Affecting Takeoff Distance

We need to consider four variables that affect takeoffs:

1) Density
2) Wind
3) Slope
4) Technique

The effect of density altitude on powerplant thrust depends on the type of powerplant. An increase in altitude above standard sea level will bring an immediate decrease in power output for a normally aspirated engine. However, an increase in altitude above standard sea level will not cause a decrease in power output for the turbocharged reciprocating engine until the aircraft exceeds the critical operating altitude. For those powerplants that experience a decay in thrust with an increase in altitude, the effect on the net accelerating force and acceleration rate can be approximated by assuming a direct variation with density. Actually, this assumed variation would closely approximate the effect on aircraft with high thrust-to-weight ratios. Density can be summarized by temperature, pressure, and moisture.

A high temperature means there is a low density. A low pressure means there is a low density. High humidity means there is a low density.

The wind issue is addressed directly in the aircraft performance charts made for the aircraft. It should be noted that the aircraft starts the takeoff roll at the velocity of the headwind. This can be determined by using the formula:

$$Takeoff\ distance = \left(\frac{Vlo - Vhw}{Vlo}\right)2$$

Proper accounting of pressure altitude and temperature is mandatory for accurate prediction of takeoff roll distance. The most critical conditions of takeoff performance are the result of some combination of high gross weight, altitude, temperature, and unfavorable wind. In all cases, the pilot must make an accurate prediction of takeoff distance from the performance data of the Airplane Flight Manual/Pilot's Operating Handbook (AFM/POH), regardless of the runway available, and strive for a polished, professional takeoff procedure.

When calculating takeoff distance using AFM/POH data, consider the following:

- Pressure altitude and temperature—to define the effect of density altitude on distance
- Gross weight—a large effect on distance
- Wind—a large effect due to the wind or wind components along the runway
- Runway slope and condition—the effect of an incline and retarding effect of factors such as snow or ice

6.8 Landing Performance

The minimum landing distance is obtained by landing at the minimum safe speed, which allows sufficient margin above stall and provides satisfactory control and capability for a go-around. Generally, the landing speed is a fixed percentage of the stall speed or minimum control speed for the aircraft in the landing configuration. As such, the landing will be accomplished at a particular value of lift coefficient and AOA. The exact values will depend on the aircraft characteristics but, once defined, the values are independent of weight, altitude, and wind.

To obtain minimum landing distance at the specified landing speed, the forces that act on the aircraft must provide maximum deceleration during the landing roll. The forces acting on the aircraft during the landing roll may require various procedures to maintain landing deceleration at the peak value. Figure 6.33 shows the forces acting on the airplane during landing.

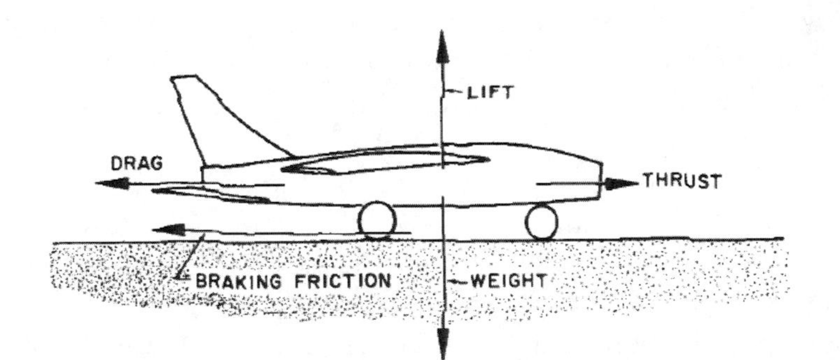

Figure 6.33. Forces Acting on the Aircraft during Landing Roll.
Source: Aerodynamics for Naval Aviators

A distinction should be made between the procedures for minimum landing distance and an ordinary landing roll with considerable excess runway available. Minimum landing distance will be obtained by creating a continuous peak deceleration of the aircraft; that is, extensive use of the brakes for maximum deceleration. On the other hand, an ordinary landing roll with considerable excess runway may allow extensive use of aerodynamic drag to minimize wear and tear on the tires and brakes. If aerodynamic drag is sufficient to cause deceleration, it can be used in deference to the brakes in the early stages of the landing roll; i.e., brakes and tires suffer from continuous hard use, but aircraft aerodynamic drag is free and does not wear out with use. The use of aerodynamic drag is applicable only for deceleration to 60 or 70 percent of the touchdown speed. At speeds less than 60 to 70 percent of the touchdown speed, aerodynamic drag is so slight as to be of little use, and braking must be utilized to produce continued deceleration. Since the objective during the landing roll is to decelerate, the powerplant thrust should be the smallest possible positive value (or largest possible negative value in aircraft equipped with thrust reversers).

Landings can be broken down into two parts, the descending approach and the ground roll. During the approach portion of the landing the airspeed is generally kept at a minimum safe approach speed of 1.3 V_{SO}. It should be noted that a higher approach speed is generally flown on final approach and during the roundout and touchdown the aircraft is slowed to 1.3 V_{SO}. Two approach angles can be used, steep or shallow. A steeper approach angle will minimize the horizontal component of inertia, reducing the flare distance. Typically the steep approach angle is used to clear obstacles in the landing path and for short field landings.

The second part of the landing to examine is the ground roll. The ground roll is considered from touchdown to 0 knots. Touchdown phase is usually at or just slightly above V_{SO}.

The forces present during the landing are the same as takeoff; thrust, drag, and friction. Again Newton's second law gives us the following formula:

$$a = \sum F/M$$

The force of the acceleration can be determined using the formula F = ma. The value of thrust at touchdown is usually 0; however, the use of reverse thrust would make this a negative value. Parasite and induced drag are present at touchdown initially, but as the aircraft slows mainly parasite drag is present.

Frictional force during landing is determined by the forces acting parallel to the runway, the landing surface, and the braking ability of the aircraft. The forces acting normal to the runway are lift and weight. Initially after touchdown the wings are still producing some lift, this means that some of the aircraft weight has not been transferred to the wheels (the wings are still supporting the aircraft); therefore the frictional force is not at its highest value. To increase the frictional force at this point some aircraft deploy spoilers to aid in this transfer. Spoilers also increase form drag and increase deceleration.

The coefficient of braking friction gives us an idea of how much of the normal forces (lift and weight) are transferred to the braking action—Figure 6.34.

Surface	μ
No Brakes	0.02
Dry Concrete	0.70
Light Rain	0. 50
Heavy Rain	0.30
Snow or Ice	0.10 - 0.20
Wheels Locked	0.50

Figure 6.34. Coefficient of Braking.

If we were to take two identical aircraft, both at different weights, and measure the ground roll on the same surface type using maximum braking we would find that the stopping distance is the same. However the heavier aircraft's brakes will be hotter because they have to dissipate more kinetic energy. This should not be confused with the fact that when we look at both parts and landing distance, approach and ground roll, the heavier aircraft will have a longer landing distance because the approach speed will be higher.

6.8.1 Minimizing Landing Distance

There are multiple factors which can minimize total landing distance. The first is to minimize V_{so}; this can be done by reducing weight, maximizing wing surface area and the coefficient of lift, and flying at lower altitudes and low temperatures (air density). Lower weights will require a slower landing speed, which will require a shorter distance to decelerate. Wing surface area can be controlled by the pilot in aircraft with a fowler type flap system. The coefficient of lift can be increased with any type of flap system. High lift devices provide the same amount of lift but at a slower airspeed. Flaps also increase the angle of descent, which reduces the horizontal component of inertia. STOL kits and vortex generators can also be used to reduce V_{so}. Flying at lower densities and temperatures will not change the actual indicated stall speed, but will reduce the TAS. This will shorten the landing distance because of the lower actual airspeed (not indicated). Another way to look at this is to look at what a high density altitude does to landing distance. An increase in density altitude increases the landing speed but does not alter the net retarding force. Thus, the aircraft at altitude lands at the same indicated airspeed (IAS) as at sea level but, because of the reduced air density, the TAS is greater. Since the aircraft lands at altitude with the same weight and dynamic pressure, the drag and braking friction throughout the landing roll have the same values as at sea level. As long as the condition is within the capability of the brakes, the net retarding force is unchanged, and the deceleration is the same as with the landing at sea level. Since an increase in altitude does not alter deceleration, the effect of high density altitude on landing distance is due to the greater TAS.

The second step in minimizing total landing distance is to maximize deceleration. This can be done by reversing the thrust vector, increasing aerodynamic drag, and increasing frictional force. Turboprops use reverse thrust by pushing the blade angle past the low pitch stop and applying engine power. In jet aircraft reverse thrust is accomplished using some form of vanes, buckets, or clamshells to turn or direct the exhaust gases forward. Friction can be increased to maximize deceleration and shorten the landing roll. Wheel braking is the predominant force in stopping the aircraft on the ground. Most high performance jets are also equipped with anti-lock brake systems to prevent the wheels from locking up. Figure 6.34 shows that the braking coefficient on dry concrete goes from .7 to .5 if the wheels are locked up. Another way to increase frictional force is to decrease lift. To get as much weight on the wheels as fast as possible spoilers are deployed after landing to destroy lift and transfer the weight faster, which will increase braking performance.

Some other factors that affect landing distance are wind, slope, and piloting technique. The effect of wind on landing distance is large and deserves proper consideration when predicting landing distance. Since the aircraft will land at a particular airspeed independent of the wind, the principal effect of wind on landing distance is the change in the groundspeed at which the aircraft touches down. The effect of wind on deceleration during the landing is identical to the effect on acceleration during the takeoff. Landing uphill will decrease the landing distance, landing downhill will increase it.

The effect of proper landing speed is important when runway lengths and landing distances are critical. The landing speeds specified in the AFM/POH are generally the minimum safe speeds at which the aircraft can be landed. Any attempt to land at below the specified speed may mean that the aircraft may stall, be difficult to control, or develop high rates of descent. On the other hand, an excessive speed at landing may improve the controllability slightly (especially in crosswinds), but causes an undesirable increase in landing distance.

A ten percent excess landing speed causes at least a 21 percent increase in landing distance. The excess speed places a greater working load on the brakes because of the additional kinetic energy to be dissipated. Also, the additional speed causes increased drag and lift in the normal ground attitude, and the increased lift reduces the normal force on the braking surfaces. The deceleration during this range of speed immediately after touchdown may suffer, and it is more probable for a tire to be blown out from braking at this point.

The most critical conditions of landing performance are combinations of high gross weight, high density altitude, and unfavorable wind. These conditions produce the greatest required landing distances and critical levels of energy dissipation required of the brakes. In all cases, it is necessary to make an accurate prediction of minimum landing distance to compare with the available runway. A polished, professional landing procedure is necessary because the landing phase of flight accounts for more pilot-caused aircraft accidents than any other single phase of flight.

In the prediction of minimum landing distance from the AFM/POH data, the following considerations must be given:

- Pressure altitude and temperature—to define the effect of density altitude
- Gross weight—which defines the CAS for landing.
- Wind—a large effect due to wind or wind components along the runway
- Runway slope and condition—relatively small correction for ordinary values of runway slope, but a significant effect of snow, ice, or soft ground

A tail wind of ten knots increases the landing distance by about 21 percent. An increase of landing speed by ten percent increases the landing distance by 20 percent. Hydroplaning makes braking ineffective until a decrease of speed to that determined using the hydroplaning formula.

Chapter 7
Stability

Stability is the inherent quality of an aircraft to correct for conditions that may disturb its equilibrium, and to return to or to continue on the original flightpath. It is primarily an aircraft design characteristic. The flightpath and attitudes an aircraft flies are limited by the aerodynamic characteristics of the aircraft, its propulsion system, and its structural strength. These limitations indicate the maximum performance and maneuverability of the aircraft. If the aircraft is to provide maximum utility, it must be safely controllable to the full extent of these limits without exceeding the pilot's strength or requiring exceptional flying ability. If an aircraft is to fly straight and steady along any arbitrary flightpath, the forces acting on it must be in static equilibrium. The reaction of any body when its equilibrium is disturbed is referred to as stability. The two types of stability are static and dynamic.

7.1 Static Stability

Static stability refers to the initial tendency, or direction of movement, back to equilibrium. In aviation, it refers to the aircraft's initial response when disturbed from a given AOA,
slip, or bank. (Figure 7.1)

Positive static stability—the initial tendency of the aircraft to return to the original state of equilibrium after being disturbed.

Neutral static stability—the initial tendency of the aircraft to remain in a new condition after its
equilibrium has been disturbed.

Negative static stability—the initial tendency of the aircraft to continue away from the original state of equilibrium after being disturbed.

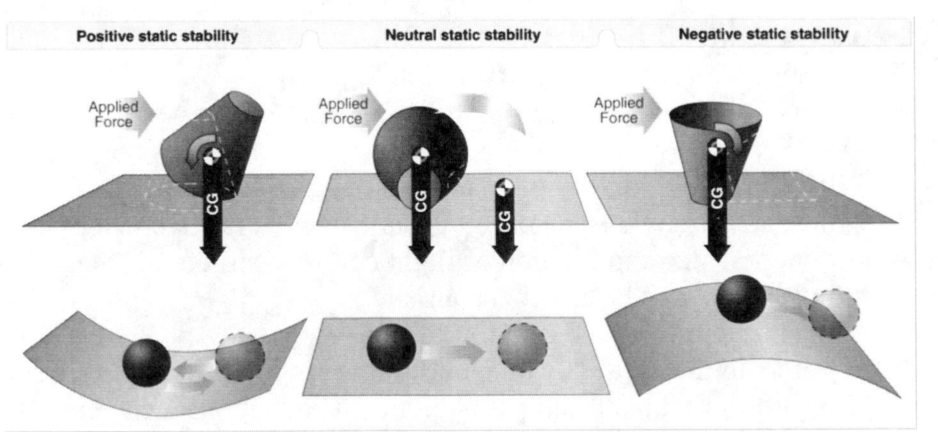

Figure 7.1. Types of Static Stability.
Source: Pilot's Handbook of Aeronautical Knowledge

7.2 Dynamic Stability

Static stability has been defined as the initial tendency to return to equilibrium that the aircraft displays after being disturbed from its trimmed condition. Occasionally, the initial tendency is different or opposite from the overall tendency, so a distinction must be made between the two.

Dynamic stability refers to the aircraft response over time when disturbed from a given AOA, slip, or bank. This type of stability also has three subtypes:

Positive dynamic stability—over time, the motion of the displaced object decreases in amplitude and, because it is positive, the object displaced returns toward the equilibrium state.
Neutral dynamic stability—once displaced, the displaced object neither decreases nor increases in amplitude. A worn automobile shock absorber exhibits this tendency.

Negative dynamic stability—over time, the motion of the displaced object increases and becomes more divergent.

Stability in an aircraft affects two areas significantly, maneuverability and controllability.

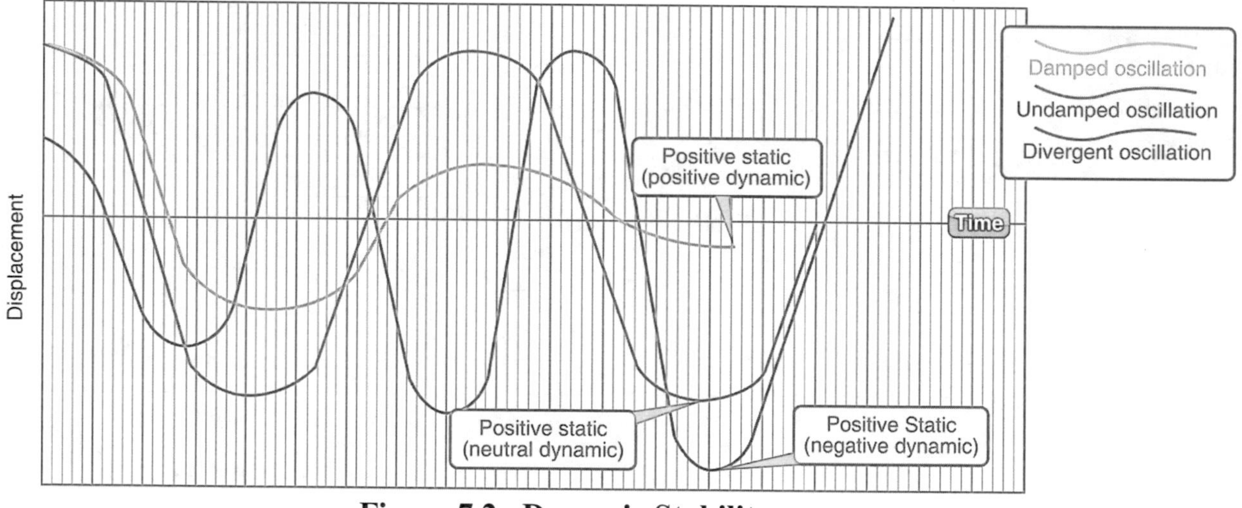

Figure 7.2. Dynamic Stability.
Source: Pilot's Handbook of Aeronautical Knowledge

Maneuverability refers to the quality of an aircraft that permits it to be maneuvered easily and to withstand the stresses imposed by maneuvers. It is governed by the aircraft's weight, inertia, size and location of flight controls, structural strength, and powerplant. It too is an aircraft design characteristic.

Controllability refers to the capability of an aircraft to respond to the pilot's control, especially with regard to flightpath and attitude. It is the quality of the aircraft's response to the pilot's control application when maneuvering the aircraft, regardless of its stability characteristics.

7.3 Longitudinal Stability (Pitching)

In designing an aircraft, a great deal of effort is spent in developing the desired degree of stability around all three axes. But longitudinal stability about the lateral axis is considered to be the most affected by certain variables in various flight conditions. **Longitudinal stability** is the quality that makes an aircraft stable about its lateral axis. (Figure 7.3) It involves the pitching motion as the aircraft's nose moves up and down in flight. A longitudinally unstable aircraft has a tendency to dive or climb progressively into a very steep dive or climb, or even a stall. Thus, an aircraft with longitudinal instability becomes difficult and sometimes dangerous to fly. Static longitudinal stability or instability in an aircraft is dependent upon three factors:

1) Location of the wing with respect to the CG
2) Location of the horizontal tail surfaces with respect to the CG
3) Area or size of the tail surfaces

In analyzing stability, it should be recalled that a body free to rotate always turns about its CG.
To obtain static longitudinal stability, the relation of the wing and tail moments must be such that, if the moments are initially balanced and the aircraft is suddenly nose up, the wing moments and tail moments change so that the sum of their forces provides an unbalanced but restoring moment which, in turn, brings the nose down again. Similarly, if the aircraft is nose down, the resulting change in moments brings the nose back up.

Figure 7.3. Longitudinal Stability.
Source: Pilot's Handbook of Aeronautical Knowledge

The C_L in most asymmetrical airfoils has a tendency to change its fore and aft positions with a change in the AOA. The C_L tends to move forward with an increase in AOA and to move aft with a decrease in AOA. This means that when the AOA of an airfoil is increased, the C_L, by moving forward, tends to lift the leading edge of the wing still more. This tendency gives the wing an inherent quality of instability. (NOTE: C_L is also known as the center of pressure [CP].) Figure 7.3 shows an aircraft in straight-and-level flight. The line CG-C_L-T represents the aircraft's longitudinal axis from the CG to a point T on the horizontal stabilizer.

Most aircraft are designed so that the wing's C_L is to the rear of the CG. This makes the aircraft "nose heavy" and requires that there be a slight downward force on the horizontal stabilizer in order to balance the aircraft and keep the nose from continually pitching downward. Compensation for this nose heaviness is provided by setting the horizontal stabilizer at a slight negative AOA. The downward force thus produced holds the tail down, counterbalancing the "heavy" nose. It is as if the line CG-C_L-T were a lever with an upward force at C_L and two downward forces balancing each other, one a strong force at the CG point and the other, a much lesser force, at point T (downward air pressure on the stabilizer). To better visualize this physics principle: If an iron bar were suspended at point C_L, with a heavy weight hanging on it at the CG, it would take downward pressure at point T to keep the "lever" in balance. Even though the horizontal stabilizer may be level when the aircraft is in level flight, there is a downwash of air from the wings. This downwash strikes the top of the stabilizer and produces a downward pressure, which at a certain speed is just enough to balance the "lever." The faster the aircraft is flying, the greater this downwash and the greater the downward force on the horizontal stabilizer (except T-tails). (Figure 7.4) In aircraft with fixed-position horizontal stabilizers, the aircraft manufacturer sets the stabilizer at an angle that provides the best stability (or balance) during flight at the design cruising speed and power setting.

If the aircraft's speed decreases, the speed of the airflow over the wing is decreased. As a result of this decreased flow of air over the wing, the downwash is reduced, causing a lesser downward force on the horizontal stabilizer. In turn, the characteristic nose heaviness is accentuated, causing the aircraft's nose to pitch down more. (Figure 7.5) This places the aircraft in a nose-low attitude, lessening the wing's AOA and drag and allowing the airspeed to increase. As the aircraft continues in the nose-low attitude and its speed increases, the downward force on the horizontal stabilizer is once again increased. Consequently, the tail is again pushed downward and the nose rises into a climbing attitude.

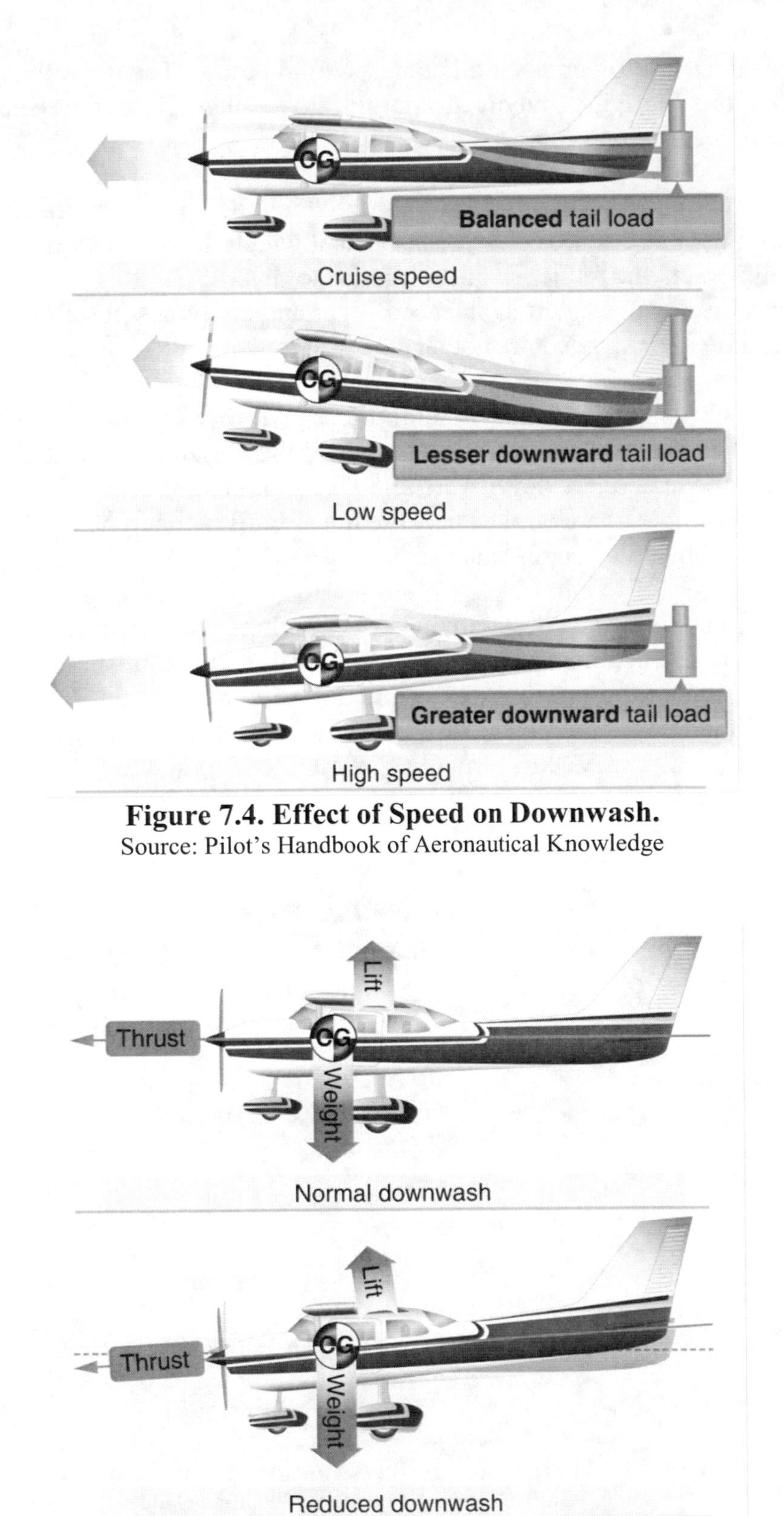

Figure 7.4. Effect of Speed on Downwash.
Source: Pilot's Handbook of Aeronautical Knowledge

Figure 7.5. Reduced Power Allows Pitch Down.
Source: Pilot's Handbook of Aeronautical Knowledge

As this climb continues, the airspeed again decreases, causing the downward force on the tail to decrease until the nose lowers once more. Because the aircraft is dynamically stable, the nose does not lower as far this time as it did before. The aircraft acquires enough speed in this more gradual dive to start it into another climb, but the climb is not as steep as the preceding one.

After several of these diminishing oscillations, in which the nose alternately rises and lowers, the aircraft finally settles down to a speed at which the downward force on the tail exactly counteracts the tendency of the aircraft to dive. When this condition is attained, the aircraft is once again in balanced flight and continues in stabilized flight as long as this attitude and airspeed are not changed.

A similar effect is noted upon closing the throttle. The downwash of the wings is reduced and the force at T in Figure 7.3 is not enough to hold the horizontal stabilizer down. It seems as if the force at T on the lever were allowing the force of gravity to pull the nose down. This is a desirable characteristic because the aircraft is inherently trying to regain airspeed and reestablish the proper balance.

Power or thrust can also have a destabilizing effect in that an increase of power may tend to make the nose rise. The aircraft designer can offset this by establishing a "high thrust line" wherein the line of thrust passes above the CG. (Figures 7.6 and 7.7) In this case, as power or thrust is increased a moment is produced to counteract the down load on the tail. On the other hand, a very "low thrust line" would tend to add to the nose-up effect of the horizontal tail surface.

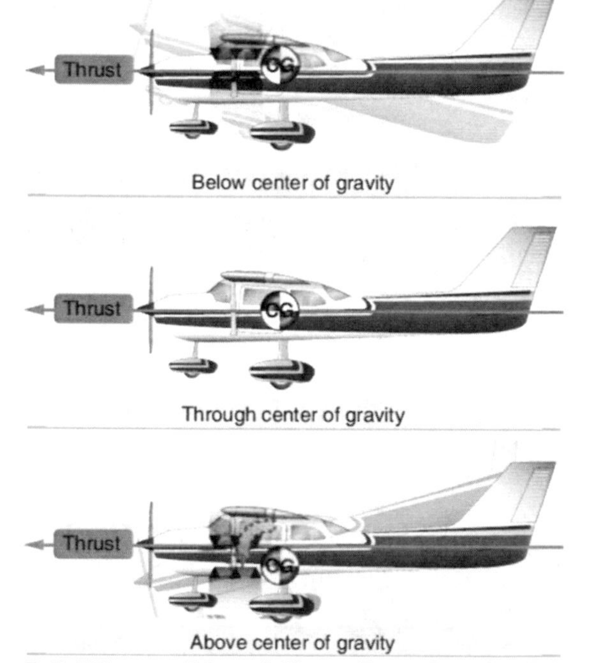

Figure 7.6. Thrust Line Affects Longitudinal Stability.
Source: Pilot's Handbook of Aeronautical Knowledge

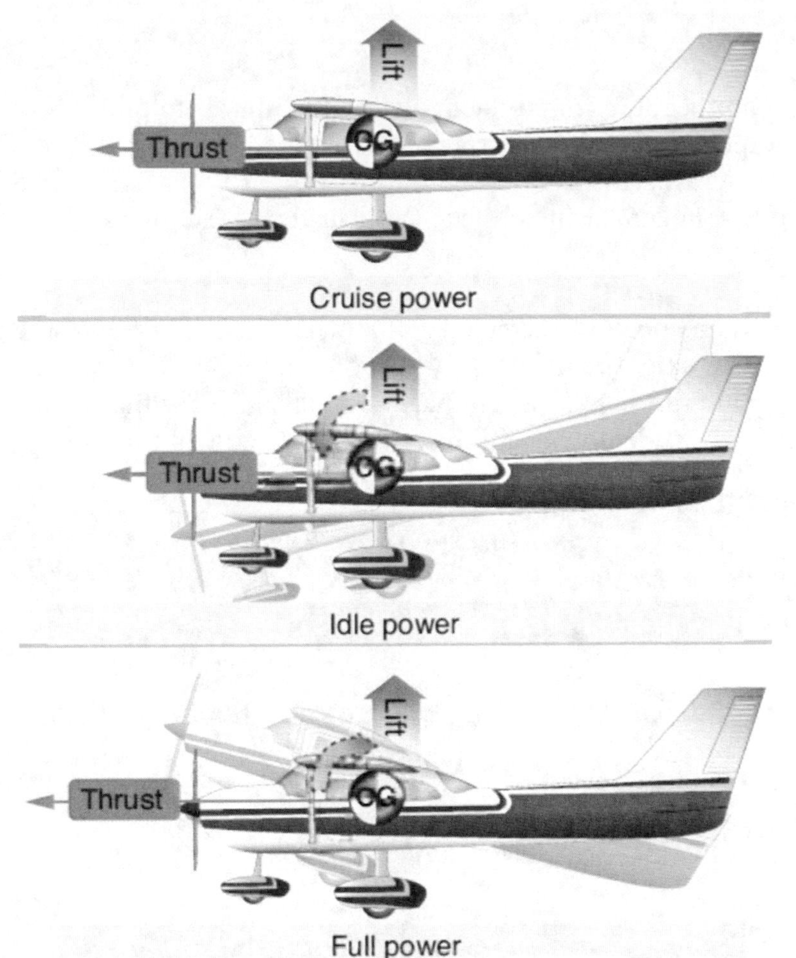

Figure 7.7. Power Changes Affect Longitudinal Stability.
Source: Pilot's Handbook of Aeronautical Knowledge

Conclusion: with CG forward of the CL and with an aerodynamic tail-down force, the aircraft usually tries to return to a safe flying attitude. The following is a simple demonstration of longitudinal stability. Trim the aircraft for "hands off" control in level flight. Then, momentarily give the controls a slight push to nose the aircraft down. If, within a brief period, the nose rises to the original position and then stops, the aircraft is statically stable. Ordinarily, the nose passes the original position (that of level flight) and a series of slow pitching oscillations follows. If the oscillations gradually cease, the aircraft has positive stability; if they continue unevenly, the aircraft has neutral stability; if they increase, the aircraft is unstable.

7.4 Lateral Stability (Rolling)

Stability about the aircraft's longitudinal axis, which extends from the nose of the aircraft to its tail, is called **lateral stability**. This helps to stabilize the lateral or "rolling effect" when one wing gets lower than the wing on the opposite side of the aircraft. There are four main design factors that make an aircraft laterally stable: dihedral, sweepback, keel effect, and weight distribution.

7.4.1 Dihedral

The most common procedure for producing lateral stability is to build the wings with an angle of one to three degrees above perpendicular to the longitudinal axis. The wings on either side of the aircraft join the fuselage to form a slight V or angle called "**dihedral**." The amount of dihedral is measured by the angle made by each wing above a line parallel to the lateral axis.

Dihedral involves a balance of lift created by the wings' AOA on each side of the aircraft's longitudinal axis. If a momentary gust of wind forces one wing to rise and the other to lower, the aircraft banks. When the aircraft is banked without turning, the tendency to sideslip or slide downward toward the lowered wing occurs. (Figure 7.8)

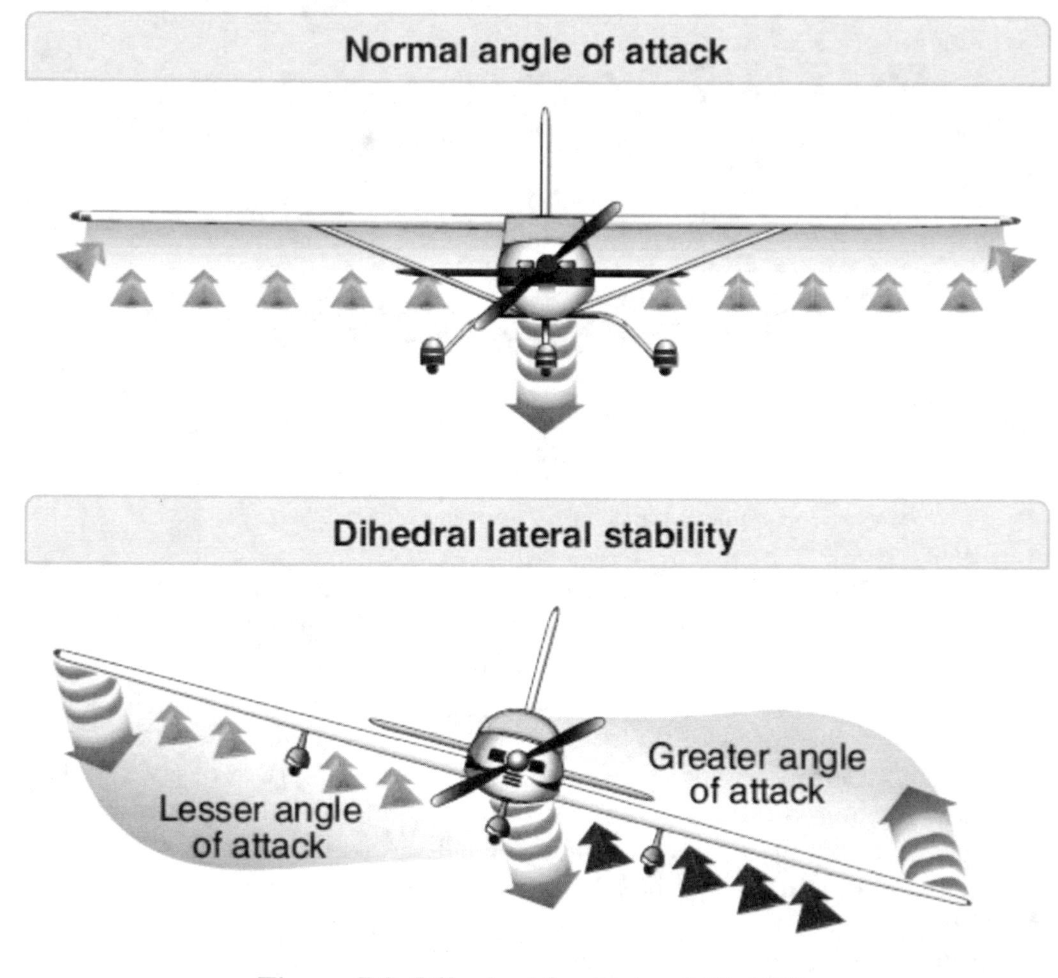

Figure 7.8. Dihedral for Lateral Stability.
Source: Pilot's Handbook of Aeronautical Knowledge

Since the wings have dihedral, the air strikes the lower wing at a much greater AOA than the higher wing. The increased AOA on the lower wing creates more lift than the higher wing. Increased lift causes the lower wing to begin to rise upward. As the wings approach the level position, the AOA on both wings once again are equal, causing the rolling tendency to subside. The effect of dihedral is to produce a rolling tendency to return the aircraft to a laterally balanced flight condition when a sideslip occurs. The restoring force may move the low wing up too far, so that the opposite wing now goes down. If so, the process is repeated, decreasing with each lateral oscillation until a balance for wings-level flight is finally reached.

7.4.2 Sweepback

Sweepback is an addition to the dihedral that increases the lift created when a wing drops from the level position. A sweptback wing is one in which the leading edge slopes backward. When a disturbance causes an aircraft with sweepback to slip or drop a wing, the low wing presents its leading edge at an angle that is perpendicular to the relative airflow. As a result, the low wing acquires more lift, rises, and the aircraft is restored to its original flight attitude. Sweepback also contributes to directional stability. When turbulence or rudder application causes the aircraft to yaw to one side, the right wing presents a longer leading edge perpendicular to the relative airflow. The airspeed of the right wing increases and it acquires more drag than the left wing. The additional drag on the right wing pulls it back, turning the aircraft back to its original path.

7.4.3 Keel Effect and Weight Distribution

An aircraft always has the tendency to turn the longitudinal axis of the aircraft into the relative wind. This "weather vane" tendency is similar to the keel of a ship and exerts a steadying influence on the aircraft laterally about the longitudinal axis. When the aircraft is disturbed and one wing dips, the fuselage weight acts like a pendulum, returning the airplane to its original attitude. (Figure 7.10)

7.5 Vertical Stability (Yawing)

Stability about the aircraft's vertical axis (the sideways moment) is called **yawing** or directional stability. Yawing or directional stability is the most easily achieved stability in aircraft design. The area of the vertical fin and the sides of the fuselage aft of the CG are the prime contributors which make the aircraft act like the well-known weather vane or arrow, pointing its nose into the relative wind.

In examining a weather vane, it can be seen that if exactly the same amount of surface were exposed to the wind in front of the pivot point as behind it, the forces fore and aft would be in balance and little or no directional movement would result. Consequently, it is necessary to have a greater surface aft of the pivot point than forward of it.

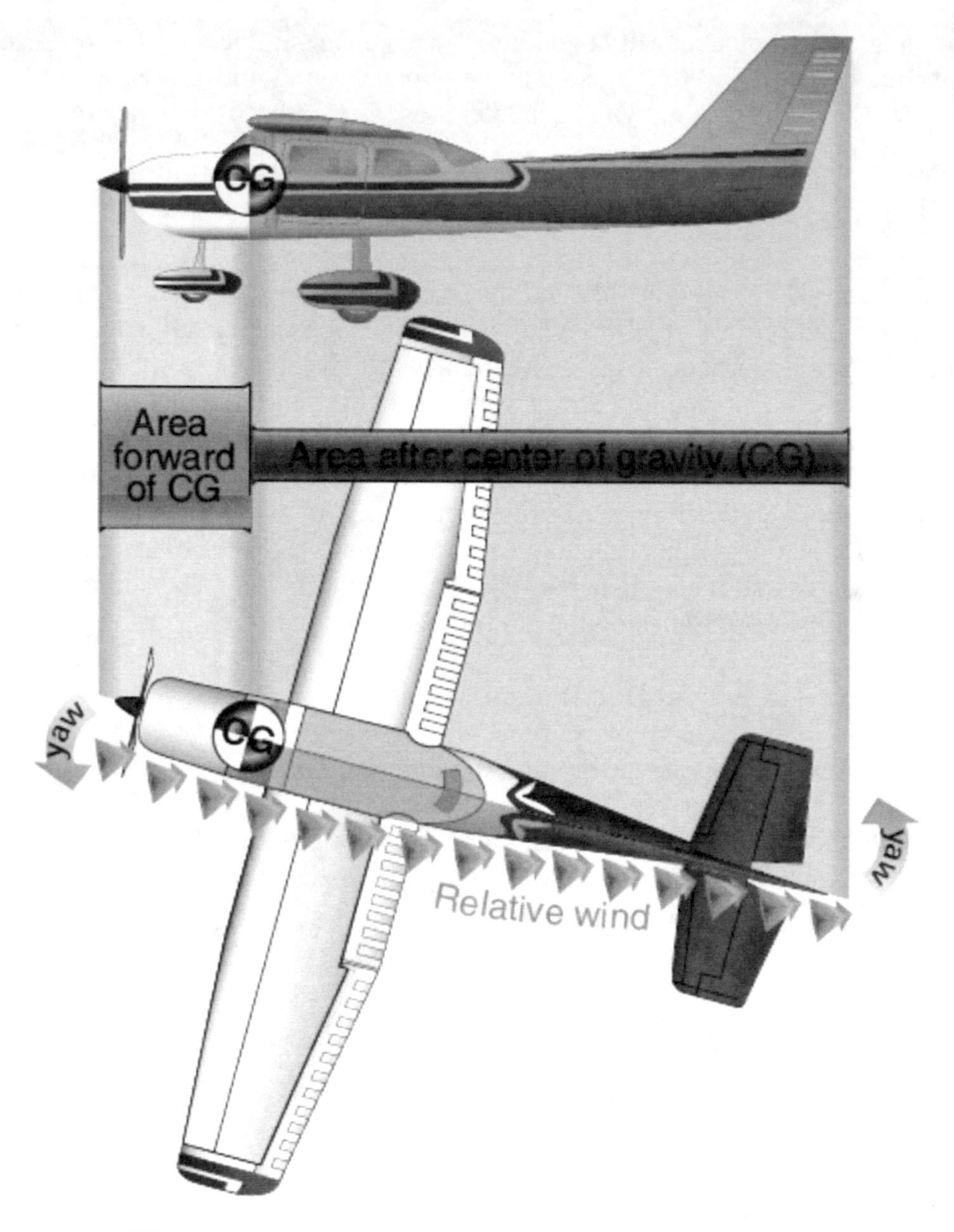

Figure 7.9. Fuselage and Fin for Lateral Stability.
Source: Pilot's Handbook of Aeronautical Knowledge

Similarly, the aircraft designer must ensure positive directional stability by making the side surface greater aft than ahead of the CG. (Figure 7.9) To provide additional positive stability to that provided by the fuselage, a vertical fin is added. The fin acts similar to the feather on an arrow in maintaining straight flight. Like the weather vane and the arrow, the farther aft this fin is placed and the larger its size, the greater the aircraft's directional stability.

If an aircraft is flying in a straight line, and a sideward gust of air gives the aircraft a slight rotation about its vertical axis (e.g., the right), the motion is retarded and stopped by the fin because while the aircraft is rotating to the right, the air is striking the left side of the fin at an angle. This causes pressure on the left side of the fn, which resists the turning motion and slows down the aircraft's yaw. In doing so, it acts somewhat like the weather vane by turning the aircraft into the relative wind. The initial change in direction of the aircraft's flightpath is generally slightly behind its change of heading. Therefore, after a slight yawing of the aircraft to the right, there is a brief moment when the aircraft is still moving along its original path, but its longitudinal axis is pointed slightly to the right. The aircraft is then momentarily skidding sideways, and during that moment (since it is assumed that although the yawing motion has stopped, the excess pressure on the left side of the fin still persists) there is necessarily a tendency for the aircraft to be turned partially back to the left. That is, there is a momentary restoring tendency caused by the fin.

This restoring tendency is relatively slow in developing and ceases when the aircraft stops skidding. When it ceases, the aircraft is flying in a direction slightly different from the original direction. In other words, it will not return of its own accord to the original heading; the pilot must reestablish the initial heading.

A minor improvement of directional stability may be obtained through sweepback. Sweepback is incorporated in the design of the wing primarily to delay the onset of compressibility during high-speed flight. In lighter and slower aircraft, sweepback aids in locating the center of pressure in the correct relationship with the CG. A longitudinally stable aircraft is built with the center of pressure aft of the CG.

Because of structural reasons, aircraft designers sometimes cannot attach the wings to the fuselage at the exact desired point. If they had to mount the wings too far forward, and at right angles to the fuselage, the center of pressure would not be far enough to the rear to result in the desired amount of longitudinal stability. By building sweepback into the wings, however, the designers can move the center of pressure toward the rear. The amount of sweepback and the position of the wings then place the center of pressure in the correct location.

The contribution of the wing to static directional stability is usually small. The swept wing provides a stable contribution depending on the amount of sweepback, but the contribution is relatively small when compared with other components.

7.5.1 Free Directional Oscillations (Dutch Roll)

Dutch roll is a coupled lateral/directional oscillation that is usually dynamically stable but is unsafe in an aircraft because of the oscillatory nature. The damping of the oscillatory mode may be weak or strong depending on the properties of the particular aircraft. If the aircraft has a right wing pushed down, the positive sideslip angle corrects the wing laterally before the nose is realigned with the relative wind. As the wing corrects the position, a lateral directional oscillation can occur resulting in the nose of the aircraft making a figure eight on the horizon as a result of two oscillations (roll and yaw), which, although of about the same magnitude, are out of phase with each other.

In most modern aircraft, except high-speed swept wing designs, these free directional oscillations usually die out automatically in very few cycles unless the air continues to be gusty or turbulent. Those aircraft with continuing Dutch roll tendencies are usually equipped with gyro-stabilized yaw dampers. Manufacturers try to reach a midpoint between too much and too little directional stability. Because it is more desirable for the aircraft to have "spiral instability" than Dutch roll tendencies, most aircraft are designed with that characteristic.

7.5.2 Spiral Instability

Spiral instability exists when the static directional stability of the aircraft is very strong as compared to the effect of its dihedral in maintaining lateral equilibrium. When the lateral equilibrium of the aircraft is disturbed by a gust of air and a sideslip is introduced, the strong directional stability tends to yaw the nose into the resultant relative wind while the comparatively weak dihedral lags in restoring the lateral balance. Due to this yaw, the wing on the outside of the turning moment travels forward faster than the inside wing and, as a consequence, its lift becomes greater. This produces an overbanking tendency which, if not corrected by the pilot,
results in the bank angle becoming steeper and steeper. At the same time, the strong directional stability that yaws the aircraft into the relative wind is actually forcing the nose to a lower pitch attitude. A slow downward spiral begins which, if not counteracted by the pilot, gradually increases into a steep spiral dive. Usually the rate of divergence in the spiral motion is so gradual that the pilot can control the tendency without any difficulty.

All aircraft are affected to some degree by this characteristic, although they may be inherently stable in all other normal parameters. This tendency explains why an aircraft cannot be flown "hands off" indefinitely.

Much research has gone into the development of control devices (wing leveler) to correct or eliminate this instability. The pilot must be careful in application of recovery controls during advanced stages of this spiral condition or excessive loads may be imposed on the structure. Improper recovery from spiral instability leading to inflight structural failures has probably contributed to more fatalities in general aviation aircraft than any other factor. Since the airspeed in the spiral condition builds up rapidly, the application of back elevator force to reduce this speed and to pull the nose up only "tightens the turn," increasing the load factor. The results
of the prolonged uncontrolled spiral are inflight structural failure or crashing into the ground, or both. The most common recorded causes for pilots who get into this situation are: loss of horizon reference, inability to control the aircraft by reference to instruments, or a combination of both.

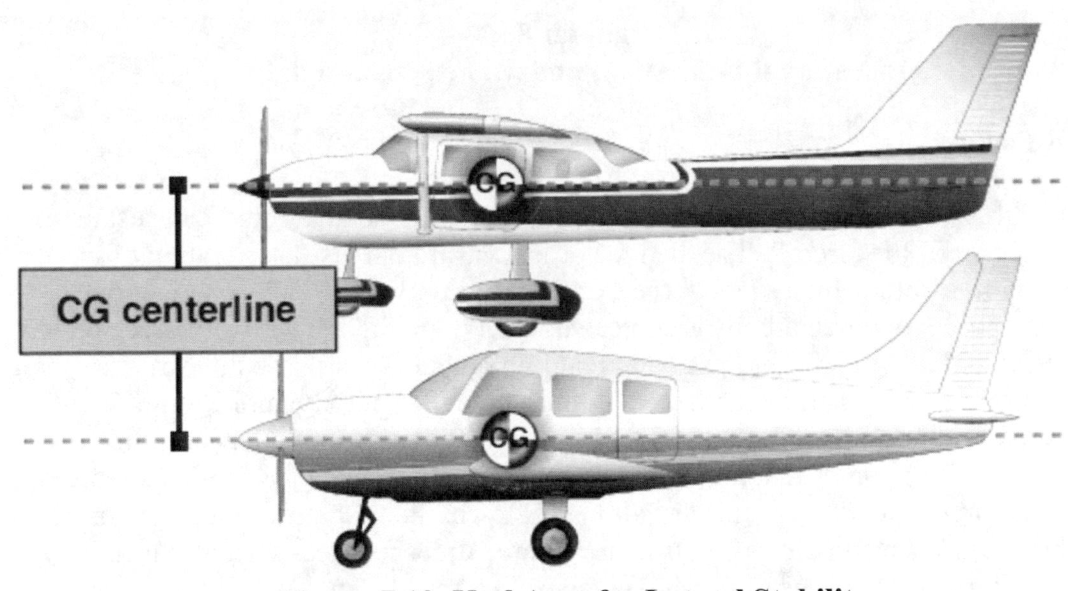

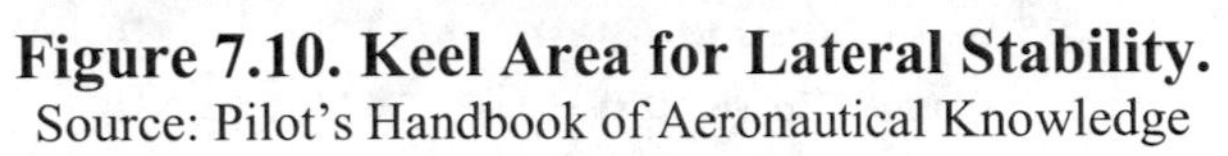

Figure 7.10. Keel Area for Lateral Stability.
Source: Pilot's Handbook of Aeronautical Knowledge

Chapter 8
Operational Limitations and Maneuvering Flight

8.1 Load Factors

In aerodynamics, **load factor** is the ratio of the maximum load an aircraft can sustain to the gross weight of the aircraft. The load factor is measured in Gs (acceleration of gravity), a unit of force equal to the force exerted by gravity on a body at rest, and indicates the force to which a body is subjected when it is accelerated. Any force applied to an aircraft to deflect its flight from a straight line produces a stress on its structure, and the amount of this force is the load factor. While a course in aerodynamics is not a prerequisite for obtaining a pilot's license, the competent pilot should have a solid understanding of the forces that act on the aircraft, the advantageous use of these forces, and the operating limitations of the aircraft being flown. For example, a load factor of 3 means the total load on an aircraft's structure is three times its gross weight. Since load factors are expressed in terms of Gs, a load factor of 3 may be spoken of as 3 Gs, or a load factor of 4 as 4 Gs. If an aircraft is pulled up from a dive, subjecting the pilot to 3 Gs, he or she would be pressed down into the seat with a force equal to three times his or her weight. Since modern aircraft operate at significantly higher speeds than older aircraft, increasing the magnitude of the load factor, this effect has become a primary consideration in the design of the structure of all aircraft. With the structural design of aircraft planned to withstand only a certain amount of overload, a knowledge of load factors has become essential for all pilots. Load factors are important for two reasons:

- It is possible for a pilot to impose a dangerous overload on the aircraft structures.
- An increased load factor increases the stalling speed and makes stalls possible at seemingly safe flight speeds.

8.2 Load Factors in Aircraft Design

The answer to the question "How strong should an aircraft be?" is determined largely by the use to which the aircraft is subjected. This is a difficult problem because the maximum possible loads are much too high for use in efficient design. It is true that any pilot can make a very hard landing or an extremely sharp pull up from a dive, which would result in abnormal loads. However, such extremely abnormal loads must be dismissed somewhat if aircraft are built that take off quickly, land slowly, and carry worthwhile payloads. The problem of load factors in aircraft design becomes how to determine the highest load factors that can be expected in normal operation under various operational situations. These load factors are called "**limit load factors**." For reasons of safety, it is required that the aircraft be designed to withstand these load factors without any structural damage. Although the Code of Federal Regulations (CFR) requires that the aircraft structure be capable of supporting one and one-half times these limit load factors without failure, it is accepted that parts of the aircraft may bend or twist under these loads and that some structural damage may occur. This 1.5 load limit factor is called the "**factor of safety**" and provides, to some extent, for loads higher than those expected under normal and reasonable operation. This strength reserve is not something which pilots should willfully abuse; rather, it is there for protection when encountering unexpected conditions. The above considerations apply to all loading conditions, whether they be due to gusts, maneuvers, or landings. The gust load factor requirements now in effect are substantially the same as those that have been in existence for years. Hundreds of thousands of operational hours have proven them adequate for safety. Since the pilot has little control over gust load factors (except to reduce the aircraft's speed when rough air is encountered), the gust loading requirements are substantially the same for most general aviation type aircraft regardless of their operational use. Generally, the gust load factors control the design of aircraft which are intended for strictly non-acrobatic usage. An entirely different situation exists in aircraft designed with maneuvering load factors. It is necessary to discuss this matter separately with respect to: 1) aircraft designed in accordance with the category system (i.e., normal, utility, acrobatic); and 2) older designs built according to requirements which did not provide for operational categories. Aircraft designed under the category system are readily identified by a placard in the flight deck, which states the operational category (or categories) in which the aircraft is certificated. The maximum safe load factors (limit load factors) specified for aircraft in the various categories are:

CATEGORY LIMIT LOAD FACTORS

Normal	1 3.8 to −1.52
Utility (mild acrobatics, including spins)	4.4 to −1.76
Acrobatic	6.0 to −3.00

- For aircraft with gross weight of more than 4,000 pounds, the limit load factor is reduced.
- To the limit loads given above, a safety factor of 50 percent is added.

There is an upward graduation in load factor with the increasing severity of maneuvers. The category system provides for maximum utility of an aircraft. If normal operation alone is intended, the required load factor (and consequently the weight of the aircraft) is less than if the aircraft is to be employed in training or acrobatic maneuvers as they result in higher maneuvering loads. Aircraft that do not have the category placard are designs that were constructed under earlier engineering requirements in which no operational restrictions were specifically given to the pilots. For aircraft of this type (up to weights of about 4,000 pounds), the required strength is comparable to present day utility category aircraft, and the same types of operation are permissible. For aircraft of this type over 4,000 pounds, the load factors decrease with weight. These aircraft should be regarded as being comparable to the normal category aircraft designed under the category system, and they should be operated accordingly.

8.3 Load Factors in Steep Turns

In a constant altitude, coordinated turn in any aircraft, the load factor is the result of two forces: centrifugal force and gravity. (Figure 8.1) For any given bank angle, the rate of turn (ROT; see Section 8.12) varies with the airspeed—the higher the speed, the slower the ROT. This compensates for added centrifugal force, allowing the load factor to remain the same.

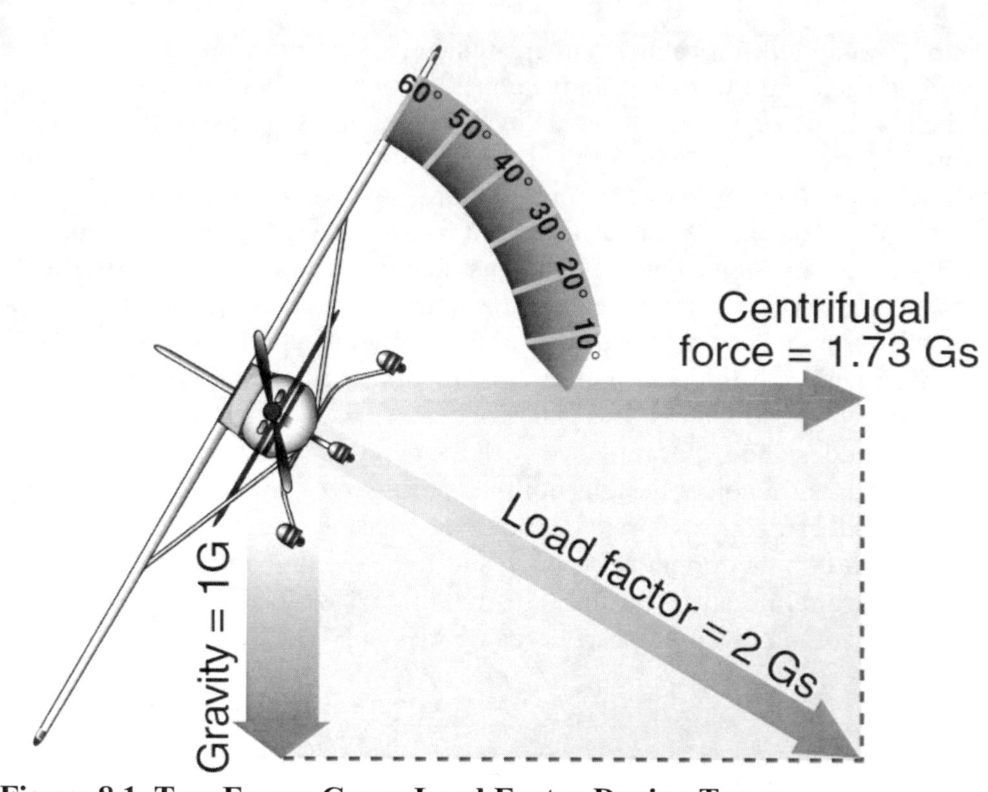

Figure 8.1. Two Forces Cause Load Factor During Turns.
Source: Pilot's Handbook of Aeronautical Knowledge

Figure 8.2 reveals an important fact about turns—the load factor increases at a terrific rate after a bank has reached 45° or 50°. The load factor for any aircraft in a 60° bank is 2 Gs. The load factor in an 80° bank is 5.76 Gs. The wing must produce lift equal to these load factors if altitude is to be maintained.

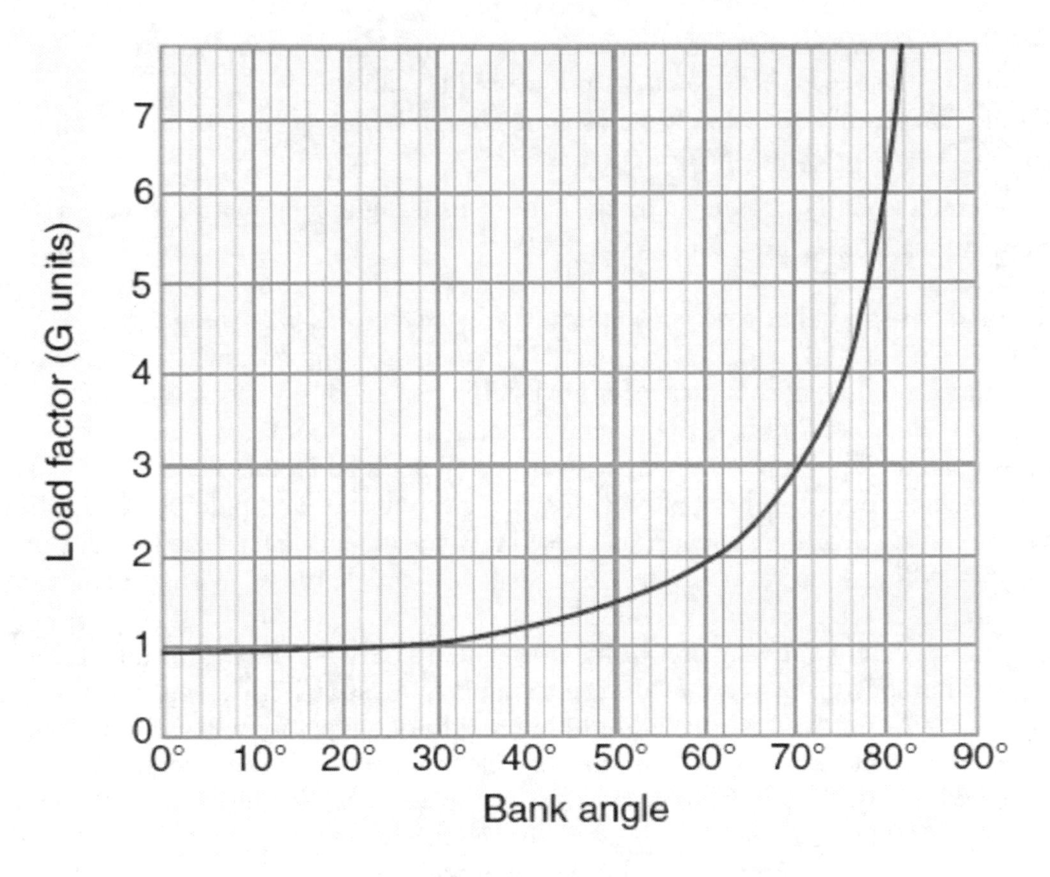

Figure 8.2. Angle of Bank Changes Load Factor.
Source: Pilot's Handbook of Aeronautical Knowledge

It should be noted how rapidly the line denoting load factor rises as it approaches the 90° bank line, which it never quite reaches because a 90° banked, constant altitude turn is not mathematically possible. An aircraft may be banked to 90°, but not in a coordinated turn. An aircraft which can be held in a 90° banked slipping turn is capable of straight knife-edged flight. At slightly more than 80°, the load factor exceeds the limit of 6 Gs, the limit load factor of an acrobatic aircraft.

For a coordinated, constant altitude turn, the approximate maximum bank for the average general aviation aircraft is 60°. This bank and its resultant necessary power setting reach the limit of this type of aircraft. An additional 10° bank increases the load factor by approximately 1 G, bringing it close to the yield point established for these aircraft. (Figure 8.3)

8.4 Load Factors and Stalling Speeds

Any aircraft, within the limits of its structure, may be stalled ***at any airspeed***. When a sufficiently high AOA is imposed, the smooth flow of air over an airfoil breaks up and separates, producing an abrupt change of flight characteristics and a sudden loss of lift, which results in a stall.

A study of this effect has revealed that the aircraft's stalling speed increases in proportion to the square root of the load factor. This means that an aircraft with a normal unaccelerated stalling speed of 50 knots can be stalled at 100 knots by inducing a load factor of 4 Gs. If it were possible for this aircraft to withstand a load factor of nine, it could be stalled at a speed of 150 knots. A pilot should be aware:

- Of the danger of inadvertently stalling the aircraft by increasing the load factor, as in a steep turn or spiral;
- When intentionally stalling an aircraft above its design maneuvering speed, a tremendous load factor is imposed.

Figures 8.2 and 8.3 show that banking an aircraft greater than 72° in a steep turn produces a load factor of 3, and the stalling speed is increased significantly. If this turn is made in an aircraft with a normal unaccelerated stalling speed of 45 knots, the airspeed must be kept greater than 75 knots to prevent inducing a stall. A similar effect is experienced in a quick pull up, or any maneuver producing load factors above 1 G. This sudden, unexpected loss of control, particularly in a steep turn or abrupt application of the back elevator control near the ground, has caused many accidents. Since the load factor is squared as the stalling speed doubles, tremendous loads may be imposed on structures by stalling an aircraft at relatively high airspeeds.

The maximum speed at which an aircraft may be stalled safely is now determined for all new designs. This speed is called the "**design maneuvering speed**" (V_A) and must be entered in the FAA-approved Airplane Flight Manual/Pilot's Operating Handbook (AFM/POH) of all recently designed aircraft. For older general aviation aircraft, this speed is approximately 1.7 times the normal stalling speed. Thus, an older aircraft which normally stalls at 60 knots must never be stalled at above 102 knots (60 knots x 1.7 = 102 knots). An aircraft with a normal stalling speed of 60 knots stalled at 102 knots undergoes a load factor equal to the square of the increase in speed, or 2.89 Gs (1.7 x 1.7 = 2.89 Gs). (The above figures are approximations to be considered as a guide, and are not the exact answers to any set of problems. The design maneuvering speed should be determined from the particular aircraft's operating limitations provided by the manufacturer.)

Since the leverage in the control system varies with different aircraft (some types employ "balanced" control surfaces while others do not), the pressure exerted by the pilot on the controls cannot be accepted as an index of the load factors produced in different aircraft. In most cases, load factors can be judged by the experienced pilot from the feel of seat pressure. Load factors can also be measured by an instrument called an "accelerometer," but this instrument is not common in general aviation training aircraft. Thus, th development of the ability to judge load factors from the feel of their effect on the body is important.

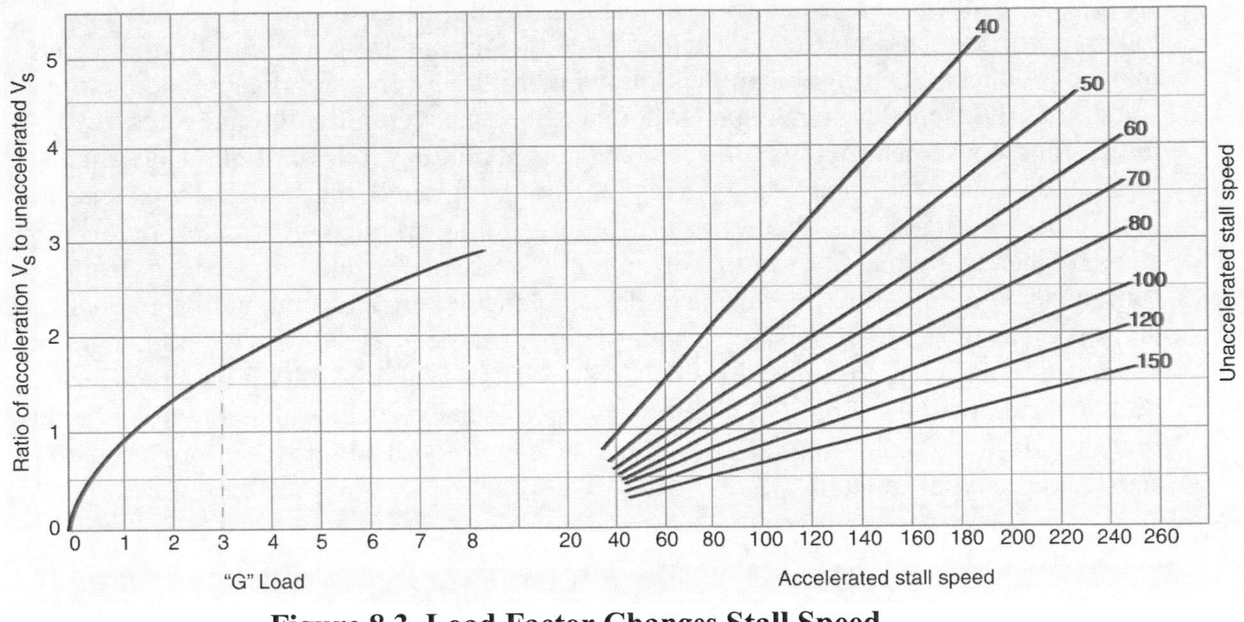

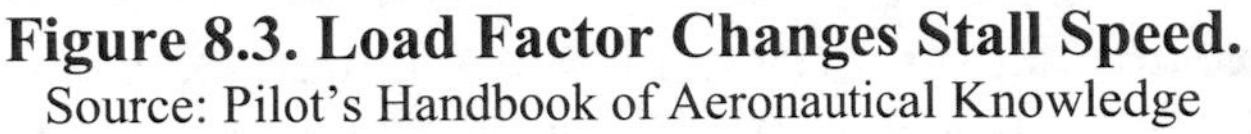

Figure 8.3. Load Factor Changes Stall Speed.
Source: Pilot's Handbook of Aeronautical Knowledge

A knowledge of these principles is essential to the development of the ability to estimate load factors. A thorough knowledge of load factors induced by varying degrees of bank and the V_A aids in the prevention of two of the most serious types of accidents:

- Stalls from steep turns or excessive maneuvering near the ground.
- Structural failures during acrobatics or other violent maneuvers resulting from loss of control.

8.5 Load Factors and Flight Maneuvers

Critical load factors apply to all flight maneuvers except unaccelerated straight flight where a load factor of 1 G is always present. Certain maneuvers considered in this section are known to involve relatively high load factors.

8.6 Turns

Increased load factors are a characteristic of all banked turns. As noted in the section on load factors in steep turns, load factors become significant to both flight performance and load on wing structure as the bank increases beyond approximately 45°. The yield factor of the average light plane is reached at a bank of approximately 70° to 75°, and the stalling speed is increased by approximately one-half at a bank of approximately 63°.

8.7 Stalls

The normal stall entered from straight-and-level flight, or an unaccelerated straight climb, does not produce added load factors beyond the 1 G of straight-and-level flight. As the stall occurs, however, this load factor may be reduced toward zero, the factor at which nothing seems to have weight. The pilot experiences a sensation of "floating free in space." If recovery is effected by snapping the elevator control forward, negative load factors (or those that impose a down load on the wings and raise the pilot from the seat) may be produced. During the pull up following stall recovery, significant load factors are sometimes induced. These may be further increased inadvertently during excessive diving (and consequently high airspeed) and abrupt pull ups to level flight. One usually leads to the other, thus increasing the load factor. Abrupt pull ups at high diving speeds may impose critical loads on aircraft structures and may produce recurrent or secondary stalls by increasing the AOA to that of stalling. As a generalization, a recovery from a stall made by diving only to cruising or design maneuvering airspeed, with a gradual pull up as soon as the airspeed is safely above stalling, can be effected with a load factor not to exceed 2 or 2.5 Gs. A higher load factor should never be necessary unless recovery has been effected with the aircraft's nose near or beyond the vertical attitude, or at extremely low altitudes to avoid diving into the ground. Chapter 9 has additional information on stalls and the types of stalls.

8.8 Spins

A stabilized spin is not different from a stall in any element other than rotation, and the same load factor considerations apply to spin recovery as apply to stall recovery. Since spin recoveries are usually effected with the nose much lower than is common in stall recoveries, higher airspeeds and consequently higher load factors are to be expected. The load factor in a proper spin recovery usually is found to be about 2.5 Gs. The load factor during a spin varies with the spin characteristics of each aircraft, but is usually found to be slightly above the 1 G of level flight. There are two reasons for this:

- Airspeed in a spin is very low, usually within 2 knots of the unaccelerated stalling speeds.
- An aircraft pivots, rather than turns, while it is in a spin.

Chapter 9 has a more detailed discussion of spins, types of spins, and spin recovery.

8.9 High Speed Stalls

The average light plane is not built to withstand the repeated application of load factors common to high speed stalls. The load factor necessary for these maneuvers produces a stress on the wings and tail structure, which does not leave a reasonable margin of safety in most light aircraft. The only way this stall can be induced at an airspeed above normal stalling involves the imposition of an added load factor, which may be accomplished by a severe pull on the elevator control. A speed of 1.7 times stalling speed (about 102 knots in a light aircraft with a stalling speed of 60 knots) produces a load factor of 3 Gs. Only a very narrow margin for error can be allowed for acrobatics in light aircraft. To illustrate how rapidly the load factor increases with airspeed, a high-speed stall at 112 knots in the same aircraft would produce a load factor of 4 Gs.

8.10 Rough Air

All standard certificated aircraft are designed to withstand loads imposed by gusts of considerable intensity. Gust load factors increase with increasing airspeed, and the strength used for design purposes usually corresponds to the highest level flight speed. In extremely rough air, as in thunderstorms or frontal conditions, it is wise to reduce the speed to the design maneuvering speed. Regardless of the speed held, there may be gusts that can produce loads which exceed the load limits. Each specific aircraft is designed with a specific G loading that can be imposed on the aircraft without causing structural damage. There are two types of load factors factored into aircraft design, limit load and ultimate load. The **limit load** is a force applied to an aircraft that causes a bending of the aircraft structure that does not return to the original shape. The **ultimate load** is the load factor applied to the aircraft beyond the limit load and at which point the aircraft material experiences structural failure (breakage). Load factors lower than the limit load can be sustained without compromising the integrity of the aircraft structure. Speeds up to but not exceeding the maneuvering speed allow an aircraft to stall prior to experiencing an increase in load factor that would exceed the limit load of the aircraft. Most AFM/POHs now include turbulent air penetration information, which helps today's pilots safely fly aircraft capable of a wide range of speeds and altitudes. It is important for the pilot to remember that the maximum "never-exceed" placard dive speeds are determined for smooth air only. High speed dives or acrobatics involving speed above the known maneuvering speed should never be practiced in rough or turbulent air.

8.11 The Vg Diagram

The flight operating strength of an aircraft is presented on a graph whose vertical scale is based on load factor. (Figure 8.4) The diagram is called a **Vg diagram**—velocity versus G loads or load factor. Each aircraft has its own Vg diagram which is valid at a certain weight and altitude. The lines of maximum lift capability (curved lines) are the first items of importance on the Vg diagram. The aircraft in Figure 8.4 is capable of developing no more than +1 G at 62 mph, the wing level stall speed of the aircraft. Since the maximum load factor varies with the square of the airspeed, the maximum positive lift capability of this aircraft is 2 G at 92 mph, 3 G at 112 mph, 4.4 G at 137 mph, and so forth. Any load factor above this line is unavailable aerodynamically (i.e., the aircraft cannot fly above the line of maximum lift capability because it stalls). The same situation exists for negative lift flight with the exception that the speed necessary to produce a given negative load factor is higher than that to produce the same positive load factor. If the aircraft is flown at a positive load factor greater than the positive limit load factor of 4.4, structural damage is possible. When the aircraft is operated in this region, objectionable permanent deformation of the primary structure may take place and a high rate of fatigue damage is incurred. Operation above the limit load factor must be avoided in normal operation. There are two other points of importance on the Vg diagram. One point is the intersection of the positive limit load factor and the line of maximum positive lift capability. The airspeed at this point is the minimum airspeed at which the limit load can be developed aerodynamically. Any airspeed greater than this provides a positive lift capability sufficient to damage the aircraft. Conversely, any airspeed less than this does not provide positive lift capability sufficient to cause damage from excessive flight loads. The usual term given to this speed is "**maneuvering speed**," since consideration of subsonic

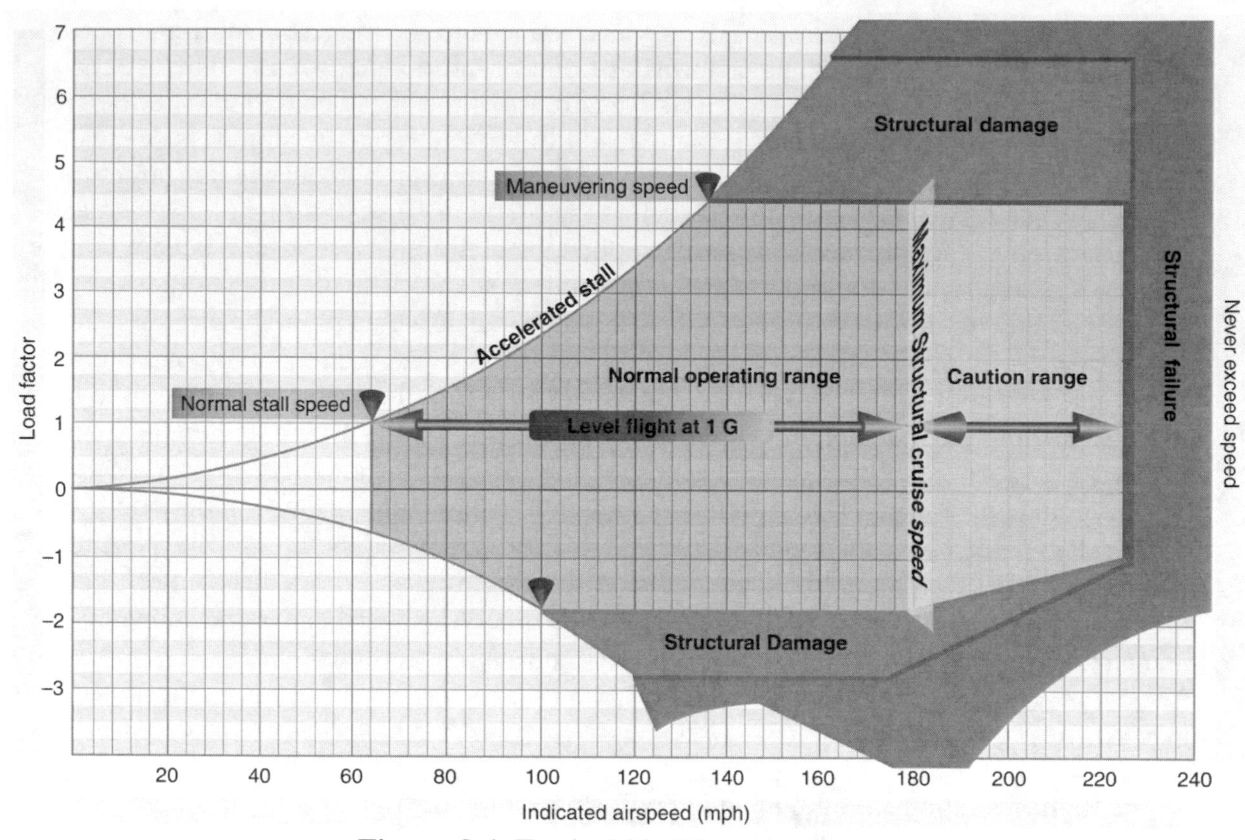

Figure 8.4. Typical Vg Diagram.
Source: Pilot's Handbook of Aeronautical Knowledge

aerodynamics would predict minimum usable turn radius or maneuverability to occur at this condition. The maneuver speed is a valuable reference point, since an aircraft operating below this point cannot produce a damaging positive flight load. Any combination of maneuver and gust cannot create damage due to excess airload when the aircraft is below the maneuver speed. The other point of importance on the Vg diagram is the intersection of the negative limit load factor and line of maximum negative lift capability. Any airspeed greater than this provides a negative lift capability sufficient to damage the aircraft; any airspeed less than this does not provide negative lift capability sufficient to damage the aircraft from excessive flight loads. The limit airspeed (or redline speed) is a design reference point for the aircraft—this aircraft is limited to 225 mph. If flight is attempted beyond the limit airspeed, structural damage or structural failure may result from a variety of phenomena. The aircraft in flight is limited to a regime of airspeeds and Gs which do not exceed the limit (or redline) speed, do not exceed the limit load factor, and cannot exceed the maximum lift capability. The aircraft must be operated within this "envelope" to prevent structural damage and ensure the anticipated service lift of the aircraft is obtained. The pilot must appreciate the Vg diagram as describing the allowable combination of airspeeds and load factors for safe operation. Any maneuver, gust, or gust plus maneuver outside the structural envelope can cause structural damage and effectively shorten the service life of the aircraft.

8.12 Rate of Turn

The **rate of turn (ROT)** is the number of degrees (expressed in degrees per second) of heading change that an aircraft makes. The ROT can be determined by taking the constant of 1,091, multiplying it by the tangent of any bank angle and dividing that product by a given airspeed in knots as illustrated in Figure 8.5.

$$\text{ROT} = \frac{1{,}091 \text{ x tangent of the bank angle}}{\text{airspeed (in knots)}}$$

Example The rate of turn for an aircraft in a coordinated turn of 30° and traveling at 120 knots would have a ROT as follows.

$$\text{ROT} = \frac{1{,}091 \text{ x tangent of } 30^\circ}{120 \text{ knots}}$$

$$\text{ROT} = \frac{1{,}091 \text{ x } 0.5773 \text{ (tangent of } 30^\circ)}{120 \text{ knots}}$$

$$\text{ROT} = 5.25 \text{ degrees per second}$$

Figure 8.5. Rate of Turn (ROT) for True Airspeed.
Source: Pilot's Handbook of Aeronautical Knowledge

If the true airspeed is increased and the ROT desired is to be constant, the angle of bank must be increased; otherwise, the ROT decreases. Likewise, if the true airspeed is held constant, an aircraft's ROT increases if the bank angle is increased. The formulas in Figures 8.5 – 8.7 depict the relationship between bank angle and true airspeed as they affect the ROT.
NOTE: All airspeed discussed in this section is true airspeed (TAS).
Airspeed significantly affects an aircraft's ROT. If airspeed is increased, the ROT is reduced if using the same angle of bank used at the lower speed.

Example Suppose we were to increase the speed to 240 knots, what is the rate of turn? Using the same formula from above we see that:

$$\text{ROT} = \frac{1{,}091 \text{ x tangent of } 30^\circ}{240 \text{ knots}}$$

ROT = 2.62 degrees per second

An increase in speed causes a decrease in the rate of turn when using the same bank angle.

Figure 8.6. An Increase in True Airspeed Causes a Decrease in the Rate of Turn.
Source: Pilot's Handbook of Aeronautical Knowledge

Therefore, if airspeed is increased as illustrated in Figure 8.6, it can be inferred that the angle of bank must be increased in order to achieve the same ROT achieved in Figure 8.7.

Example Suppose we wanted to know what bank angle would give us a rate of turn of 5.25° per second at 240 knots. A slight rearrangement of the formula would indicate it will take a 49° angle of bank to achieve the same ROT used at the lower airspeed of 120 knots.

$$\text{ROT } (5.25) = \frac{1{,}091 \text{ x tangent of X}}{240 \text{ knots}}$$

$$240 \text{ x } 5.25 = 1{,}091 \text{ x tangent of X}$$

$$\frac{240 \text{ x } 5.25}{1{,}091} = \text{tangent of X}$$

$$1.1549 = \text{tangent of X}$$

$$49^\circ = \text{X}$$

Figure 8.7. To Achieve the Same Rate of Turn of an Aircraft Traveling at 120 Knots, an Increase of Bank Angle is Required.
Source: Pilot's Handbook of Aeronautical Knowledge

What does this mean on a practicable side? If a given airspeed and bank angle produces a specific ROT, additional conclusions can be made. Knowing the ROT is a given number of degrees of change per second, the number of seconds it takes to travel 360° (a circle) can be determined by simple division. For example, if you are moving at 120 knots with a 30° bank angle, the ROT is 5.25° per second and it takes 68.6 seconds (360° divided by 5.25 = 68.6 seconds) to make a complete circle. Likewise, if you are flying at 240 knots TAS and using a 30° angle of bank, the ROT is only about 2.63° per second and it takes about 137 seconds to complete a 360° circle. Looking at the formula, any increase in airspeed is directly proportional to the time the aircraft takes to travel an arc.
So why is this important to understand? Once the ROT is understood, a pilot can determine the distance required to make that particular turn, which is explained in radius of turn (below).

8.13 Radius of Turn

The **radius of turn** is directly linked to the ROT which, as explained earlier, is a function of both bank angle and airspeed. If the bank angle is held constant and the airspeed is increased, the radius of the turn changes (increases). A higher airspeed causes the aircraft to travel through a longer arc due to a greater speed. An aircraft traveling at 120 knots is able to turn a 360° circle in a tighter radius than an aircraft traveling at 240 knots. In order to compensate for the increase in airspeed, the bank angle would need to be increased. The radius of turn (R) can be computed using a simple formula. The radius of turn is equal to the velocity squared (V^2) divided by 11.26 times the tangent of the bank angle.

$$R = \frac{V^2}{(11.26)\ (tangent\ of\ the\ bank\ angle)}$$

Using the examples provided, the turn radius for each of the two speeds can be computed. Note that if the speed is doubled, the radius is squared.

120 knots

$$R = \frac{V^2}{11.26 \times \text{tangent of bank angle}}$$

$$R = \frac{120^2}{11.26 \times \text{tangent of } 30°}$$

$$R = \frac{14{,}400}{11.26 \times 0.5773}$$

$$R = 2{,}215 \text{ feet}$$

The radius of a turn required by an aircraft traveling at 120 knots and using a bank angle of 30° is 2,215 feet.

Figure 8.8. Radius of Turn with 120 Knots True Airspeed.
Source: Pilot's Handbook of Aeronautical Knowledge

240 knots $R = \dfrac{V^2}{11.26 \times \text{tangent of bank angle}}$

$$R = \frac{240^2}{11.26 \times \text{tangent of } 30°}$$

$$R = \frac{57{,}600}{11.26 \times 0.57735}$$

$R = 8{,}861$ feet
(four times the radius at 120 knots)

The radius of a turn required by an aircraft traveling at 240 knots using the same bank angle in *Figure 4-51* is 8,861 feet. Speed is a major factor in a turn.

Figure 8.9. Radius of Turn with 240 Knots True Airspeed.
Source: Pilot's Handbook of Aeronautical Knowledge

Another way to determine the radius of turn is to use speed in feet per second (fps), π (3.1415), and the tate of turn (ROT). Using the example in Figure 8.5, it was determined that an aircraft with a ROT of 5.25 degrees per second required 68.6 seconds to make a complete circle. An aircraft's speed (in knots true airspeed [TAS]) can be converted to feet per second (fps) by multiplying it by a constant of 1.69. Therefore, an aircraft traveling at 120 knots (TAS) travels at 202.8 fps. Knowing the speed in fps (202.8) multiplied by the time an aircraft takes to complete a circle (68.6 seconds) can determine the size of the circle; 202.8 times 68.6 equals 13,912 feet. Dividing by π yields a diameter of 4,428 feet, which when divided by 2 equals a radius of 2,214 feet.

$$r = \frac{\dfrac{\text{speed (fps)} \times \dfrac{360}{\text{ROT}}}{\text{Pi } (\pi)}}{2}$$

$$r = \frac{\dfrac{202.8 \times 68.6}{\pi}}{2}$$

$$r = \frac{\dfrac{13{,}912}{\pi}}{2}$$

$$r = \frac{4{,}428}{2} = 2{,}214 \text{ feet}$$

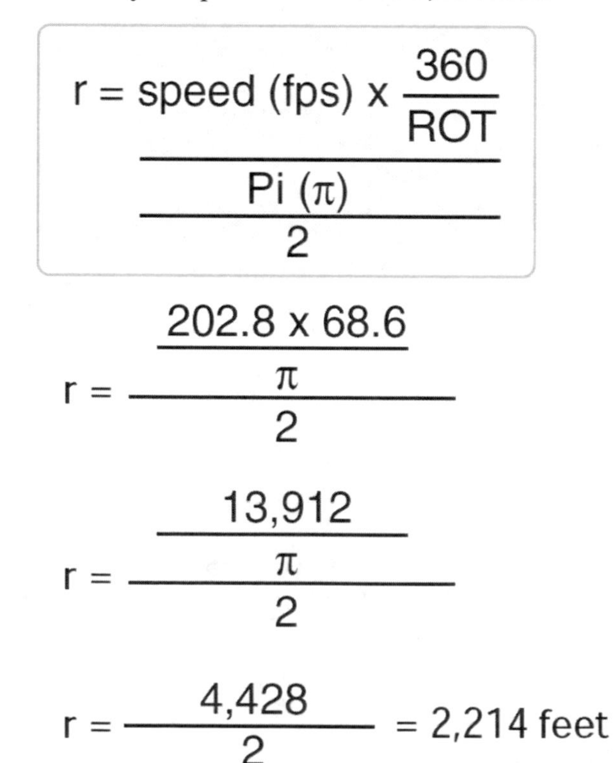

Figure 8.10. Another Formula that can be Used for Radius of Turn.
Source: Pilot's Handbook of Aeronautical Knowledge

Chapter 9
Stalls and Spins

9.0 Stalls

An aircraft **stall** results from a rapid decrease in lift caused by the separation of airflow from the wing's surface brought on by exceeding the critical AOA. A stall can occur at any pitch, attitude, or airspeed. Stalls are one of the most misunderstood areas of aerodynamics because pilots often believe that an airfoil stops producing lift when it stalls. In a stall, the wing does not totally stop producing lift. Rather, it cannot generate adequate lift to sustain level flight.

Since the CL increases with an increase in AOA, at some point the CL peaks and then begins to drop off. This peak is called the CL_{MAX}. The amount of lift the wing produces drops dramatically after exceeding the CL_{MAX} or critical AOA but, as stated above, it does not completely stop producing lift.

In most straight-wing aircraft, the wing is designed to stall the wing root first. The wing root reaches its critical AOA first, making the stall progress outward toward the wingtip. By having the wing root stall first, aileron effectiveness is maintained at the wingtips, maintaining controllability of the aircraft. Various design methods are used to achieve the stalling of the wing root first. In one design, the wing is "twisted" to a higher AOA at the wing root. Installing stall strips on the first 20 to 25 percent of the wing's leading edge is another method to introduce a stall prematurely.

The wing never completely stops producing lift in a stalled condition. If it did, the aircraft would fall to the earth. Most training aircraft are designed for the nose of the aircraft to drop during a stall, reducing the AOA and "unstalling" the wing. The "nose-down" tendency is due to the CL being aft of the CG. The CG range is very important when it comes to stall recovery characteristics. If an aircraft is allowed to be operated outside of the CG, the pilot may have difficulty recovering from a stall. The most critical CG violation would occur when operating with a CG which exceeds the rear limit. In this situation, a pilot may not be able to generate sufficient force with the elevator to counteract the excess weight aft of the CG. Without the ability to decrease the AOA, the aircraft continues in a stalled condition until it contacts the ground.

The **stalling speed** of a particular aircraft is not a fixed value for all flight situations, but a given aircraft always stalls at the same **AOA** regardless of airspeed, weight, load factor, or density altitude. Each aircraft has a particular AOA where the airflow separates from the upper surface of the wing and the stall occurs. This critical AOA varies from 16° to 20° depending on the aircraft's design. But each aircraft has only one specific AOA where the stall occurs.

There are three flight situations in which the critical AOA can be exceeded: low speed, high speed, and turning. The aircraft can be stalled in straight-and-level flight by flying too slowly. As the airspeed decreases, the AOA must be increased to retain the lift required for maintaining altitude. The lower the airspeed becomes, the more the AOA must be increased. Eventually, an AOA is reached which results in a reduced coefficient of lift. This is the **stall angle of attack**. (Figure 9.1)

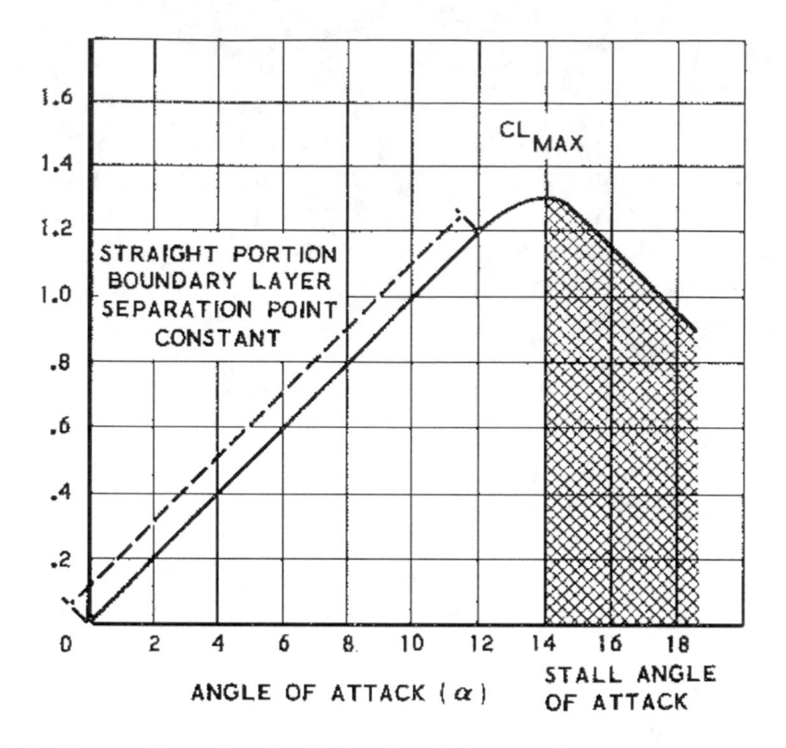

Figure 9.1. Exceeding the Stall Angle of Attack Results in a Reduced C_L.
Source: Aerodynamics for Naval Aviators

Low speed is not necessary to produce a stall. The wing can be brought into an excessive AOA at any speed. For example, an aircraft is in a dive with an airspeed of 100 knots when the pilot pulls back sharply on the elevator control. (Figure 9.2)

Figure 9.2. Forces Exerted When Pulling Out of a Dive.
Source: Pilot's Handbook of Aeronautical Knowledge

Gravity and centrifugal force prevent an immediate alteration of the flightpath, but the aircraft's AOA changes abruptly from quite low to very high. Since the flightpath of the aircraft in relation to the oncoming air determines the direction of the relative wind, the AOA is suddenly increased, and the aircraft would reach the stalling angle at a speed much greater than the normal stall speed. The stalling speed of an aircraft is also higher in a level turn than in straight-and-level flight. (Figure 9.3) Centrifugal force is added to the aircraft's weight and the wing must produce sufficient additional lift to counterbalance the load imposed by the combination of centrifugal force and weight. In a turn, the necessary additional lift is acquired by applying back pressure to the elevator control. This increases the wing's AOA, and results in increased lift. The AOA must increase as the bank angle increases to counteract the increasing load caused by centrifugal force. If at any time during a turn the AOA becomes excessive, the aircraft stalls.

At this point, the action of the aircraft during a stall should be examined. To balance the aircraft aerodynamically, the CL is normally located aft of the CG. Although this makes the aircraft inherently nose-heavy, downwash on the horizontal stabilizer counteracts this condition. At the point of stall, when the upward force of the wing's lift and the downward tail force cease, an unbalanced condition exists. This allows the aircraft to pitch down abruptly, rotating about its CG. During this nose-down attitude, the AOA decreases and the airspeed again increases. The smooth flow of air over the wing begins again, lift returns, and the aircraft is again flying. Considerable altitude may be lost before this cycle is complete.

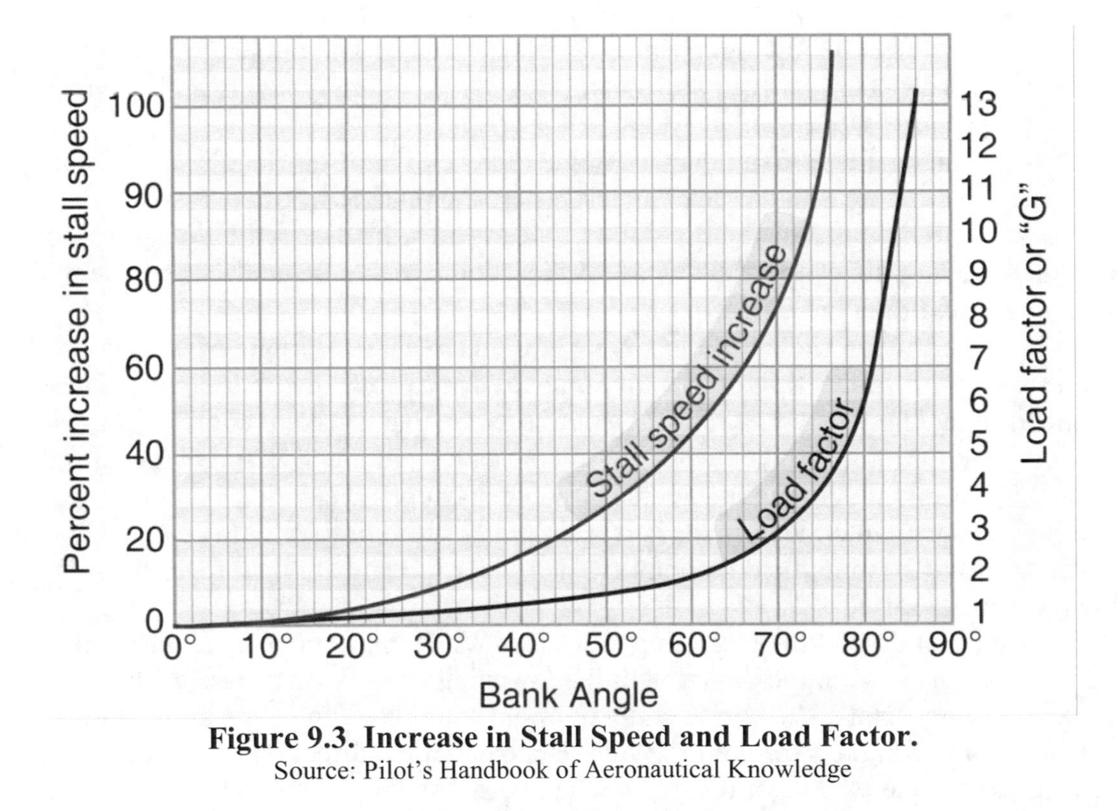

Figure 9.3. Increase in Stall Speed and Load Factor.
Source: Pilot's Handbook of Aeronautical Knowledge

Airfoil shape and degradation of that shape must also be considered in a discussion of stalls. For example, if ice, snow, and frost are allowed to accumulate on the surface of an

aircraft, the smooth airflow over the wing is disrupted. This causes the boundary layer to separate at an AOA lower than that of the critical angle. Lift is greatly reduced, altering expected aircraft performance. If ice is allowed to accumulate on the aircraft during flight, the weight of the aircraft is increased while the ability to generate lift is decreased. As little as 0.8 millimeter of ice on the upper wing surface increases drag and reduces aircraft lift by 25 percent.
Pilots can encounter icing in any season, anywhere in the country, at altitudes of up to 18,000 feet and sometimes higher. Small aircraft, including commuter planes, are most vulnerable because they fly at lower altitudes where ice is more prevalent. They also lack mechanisms common on jet aircraft that prevent ice buildup by heating the front edges of wings.
Icing can occur in clouds any time the temperature drops below freezing and super-cooled droplets build up on an aircraft and freeze. (Super-cooled droplets are still liquid even though the temperature is below 32 °F, or 0 °C).

9.1 Spins

A spin is caused when the airplane's wing exceeds its critical angle of attack (**stall**) with a sideslip or **yaw** acting on the airplane at, or beyond, the actual stall. Hence a **spin** can be defined as a stall with yaw. A spin may be further explained as an aggravated stall that results in what is termed "autorotation" wherein the airplane follows a downward corkscrew path. As the airplane rotates around a vertical axis, the rising wing is less stalled than the descending wing, creating a rolling, yawing, and pitching motion. The airplane is basically being forced downward by gravity, rolling, yawing, and pitching in a spiral path. (Figure 9.4)
The autorotation results from an unequal angle of attack on the airplane's wings. The rising wing has a decreasing angle of attack, where the relative lift increases and the drag decreases. In effect, this wing is less stalled. Meanwhile, the descending wing has an increasing angle of attack, past the wing's critical angle of attack (stall) where the relative lift decreases and drag increases.
A spin is caused when the airplane's wing exceeds its critical angle of attack (stall) with a sideslip or yaw acting on the airplane at, or beyond, the actual stall. During this uncoordinated maneuver, a pilot may not be aware that a critical angle of attack has been exceeded until the airplane yaws out of control toward the lowering wing. If stall recovery is not initiated immediately, the airplane may enter a spin.
If this stall occurs while the airplane is in a slipping or skidding turn, this can result in a spin entry and rotation in the direction that the rudder is being applied, regardless of which wingtip is raised.
Often a wing will drop at the beginning of a stall. When this happens, the nose will attempt to move (yaw) in the direction of the low wing. This is where use of the rudder is important during a stall. The correct amount of opposite rudder must be applied to keep the nose from yawing toward the low wing. By maintaining directional control and not allowing the nose to yaw toward the low wing, before stall recovery is initiated, a spin will be averted. If the nose is allowed to yaw during the stall, the airplane will begin to slip in the direction of the lowered wing, and will enter a spin.

An airplane must be stalled in order to enter a spin; therefore, continued practice in stalls will help the pilot develop a more instinctive and prompt reaction in recognizing an approaching spin. It is essential to learn to apply immediate corrective action any time it is apparent that the airplane is nearing spin conditions. If it is impossible to avoid a spin, the pilot should immediately execute spin recovery procedures.

The primary cause of an inadvertent spin is exceeding the critical AOA while applying excessive or insufficient rudder and, to a lesser extent, aileron. Insufficient or excessive control inputs to correct for **power factor (PF)**, or asymmetric propeller loading, could aggravate the precipitation of a spin. At a high AOA the downward moving blade, which is normally on the right side of the propeller arc, has a higher AOA and therefore higher thrust than the upward moving blade on the left. This results in a tendency for the airplane to yaw around the vertical axis to the left. If insufficient or excessive rudder correction is applied to counteract PF, uncoordinated flight may result. A classic situation where PF could play an important role in a stall/spin accident is during a go-around or short field takeoff where the airplane is at a high pitch attitude, high power setting, and low airspeed. In an uncoordinated maneuver, the pitot/static instruments, especially the altimeter and airspeed indicator, are unreliable due to the uneven distribution of air pressure over the fuselage. The pilot may not be aware that a critical AOA is approaching until the stall warning device activates. If a stall recovery is not promptly initiated, the airplane is more likely to enter an inadvertent spin. For example, stall/spin accidents have occurred during a turn from base to final because the pilot attempted to rudder the airplane around (skid) so as not to overshoot the runway nor use excessive bank angle in the traffic pattern. The spin that occurs from cross controlling an aircraft usually results in rotation in the direction of the rudder being applied, regardless of which wingtip is raised. In a skidding turn, where both aileron and rudder are applied in the same direction, rotation will be in the direction the controls are applied. However, in a slipping turn, where opposite aileron is held against the rudder, the resultant spin will usually occur in the direction opposite the aileron that is being applied.

9.2 The Four Phases of a Spin

There are four phases of a spin: entry, incipient, developed, and recovery. (Figure 9.4)

9.2.1 Entry Phase

The **entry phase** is where the pilot provides the necessary elements for the spin, either accidentally or intentionally. The entry procedure for demonstrating a spin is similar to a power-off stall. During the entry, the power should be reduced slowly to idle, while simultaneously raising the nose to a pitch attitude that will ensure a stall. As the airplane approaches a stall, smoothly apply full rudder in the direction of the desired spin rotation while applying full back (up) elevator to the limit of travel. Always maintain the ailerons in the neutral position during the spin procedure unless the AFM/POH specifies otherwise.

9.2.2 Incipient Phase

The **incipient phase** is from the time the airplane stalls and rotation starts until the spin has fully developed. This change may take up to two turns for most airplanes. Incipient spins that are not allowed to develop into a steady-state spin are the most commonly used in the introduction to spin training and recovery techniques. In the incipient phase, the aerodynamic and inertial forces have not achieved a balance. As the incipient spin develops, the indicated airspeed should be near or below stall airspeed, and the turn-and-slip indicator should indicate the direction of the spin. The incipient spin recovery procedure should be commenced prior to the completion of 360° of rotation. The pilot should apply full rudder opposite the direction of rotation. If the pilot is not sure of the direction of the spin, check the turn-and-slip indicator; it will show a deflection in the direction of rotation.

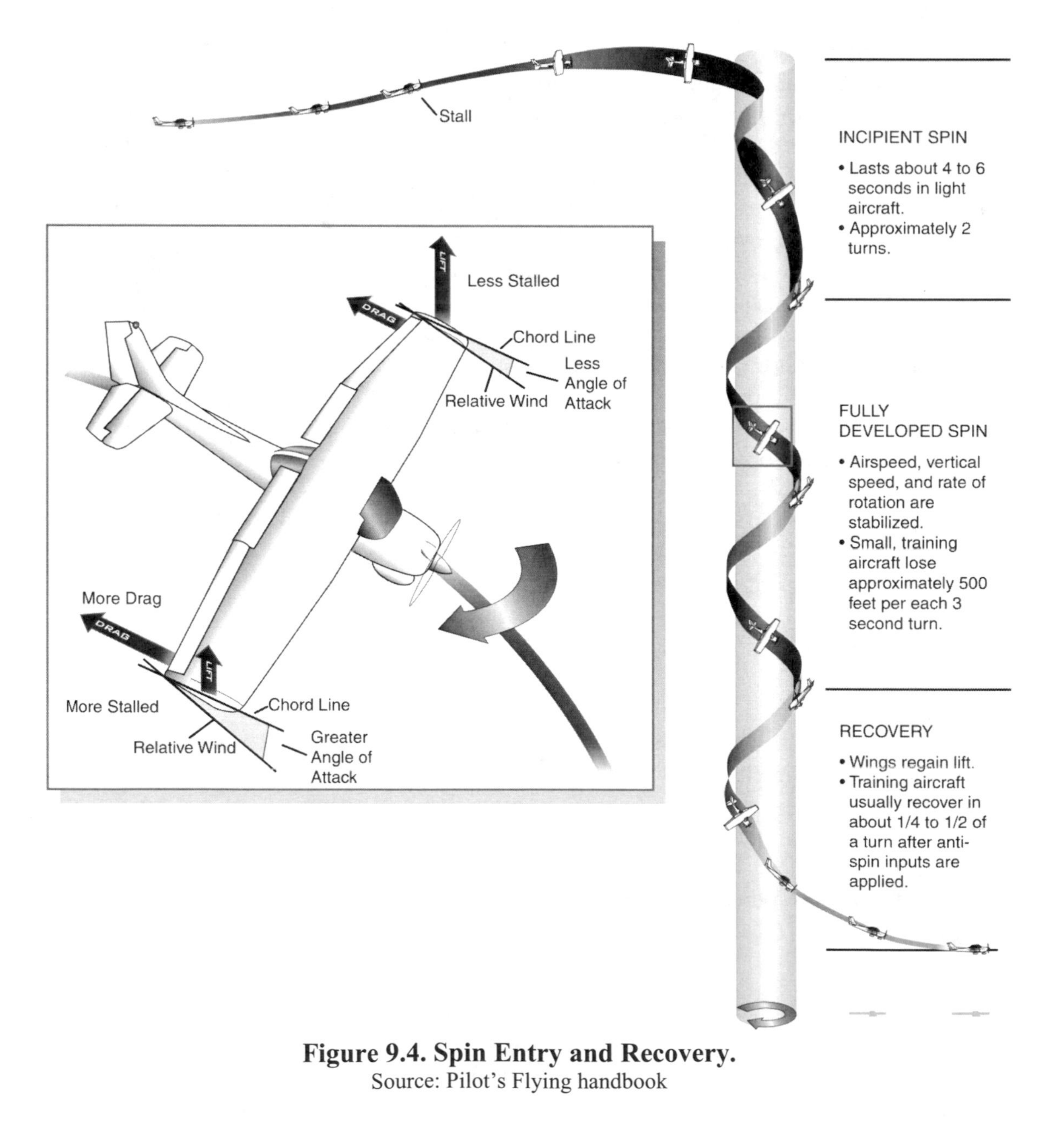

Figure 9.4. Spin Entry and Recovery.
Source: Pilot's Flying handbook

9.2.3 Developed Phase

The **developed phase** occurs when the airplane's angular rotation rate, airspeed, and vertical speed are stabilized while in a flightpath that is nearly vertical. This is where airplane aerodynamic forces and inertial forces are in balance, and the attitude, angles, and self-sustaining motions about the vertical axis are constant or repetitive. The spin is in equilibrium.
A flat spin is characterized by a near level pitch and roll attitude with the spin axis near the CG of the airplane. Recovery from a flat spin may be extremely difficult and, in some cases, impossible.

9.2.4 Recovery Phase

The **recovery phase** occurs when the angle of attack of the wings decreases below the critical angle of attack and autorotation slows. Then the nose steepens and rotation stops. This phase may last for a quarter turn to several turns.
To recover, control inputs are initiated to disrupt the spin equilibrium by stopping the rotation and stall. To accomplish spin recovery, the manufacturer's recommended procedures should be followed. In the absence of the manufacturer's recommended spin recovery procedures and techniques, the following spin recovery procedures are recommended by the FAA in *The Pilot's Flying Handbook.*
Step 1—REDUCE THE POWER (THROTTLE) TO IDLE. Power aggravates the spin characteristics. It usually results in a flatter spin attitude and increased rotation rates.
Step 2—POSITION THE AILERONS TO NEUTRAL. Ailerons may have an adverse effect on spin recovery. Aileron control in the direction of the spin may speed up the rate of rotation and delay the recovery. Aileron control opposite the direction of the spin may cause the down aileron to move the wing deeper into the stall and aggravate the situation. The best procedure is to ensure that the ailerons are neutral.
Step 3—APPLY FULL OPPOSITE RUDDER AGAINST THE ROTATION. Make sure that full (against the stop) opposite rudder has been applied.
Step 4—APPLY A POSITIVE AND BRISK, STRAIGHTFORWARD MOVEMENT OF THE ELEVATOR CONTROL FORWARD OF THE NEUTRAL TO BREAK THE STALL. This should be done immediately after full rudder application. The forceful movement of the elevator will decrease the excessive angle of attack and break the stall. The controls should be held firmly in this position. When the stall is "broken," the spinning will stop.
Step 5—AFTER SPIN ROTATION STOPS, NEUTRALIZE THE RUDDER. If the rudder is not neutralized at this time, the ensuing increased airspeed acting upon a deflected rudder will cause a yawing or skidding effect. Slow and overly cautious control movements during spin recovery must be avoided. In certain cases it has been found that such movements result in the airplane continuing to spin indefinitely, even with anti-spin inputs. A brisk and positive technique, on the other hand, results in a more positive spin recovery.

Step 6—BEGIN APPLYING BACK-ELEVATOR PRESSURE TO RAISE THE NOSE TO LEVEL FLIGHT. Caution must be used not to apply excessive back-elevator pressure after the rotation stops. Excessive back-elevator pressure can cause a secondary stall and result in another spin. Care should be taken not to exceed the "G" load limits and airspeed limitations during recovery. If the flaps and/or retractable landing gear are extended prior to the spin, they should be retracted as soon as possible after spin entry.
It is important to remember that the above spin recovery procedures and techniques are recommended for use only in the absence of the manufacturer's procedures. Before any pilot attempts to begin spin training, that pilot must be familiar with the procedures provided by the manufacturer for spin recovery.
The most common problems in spin recovery include pilot confusion as to the direction of spin rotation and whether the maneuver is a spin versus spiral. If the airspeed is increasing, the airplane is no longer in a spin but in a spiral. In a spin, the airplane is stalled. The indicated airspeed, therefore, should reflect stall speed.

9.3 Stall/Spin Avoidance Through Awareness

Remember, a spin is a result of exceeding your wing's critical angle of attack (**stall**) with a sideslip or **yaw.** Hence avoiding a stall will prevent a spin.

9.3.1 Stall Recognition

There are several ways to recognize that a stall is impending before it actually occurs. When one or more of these indicators is noted, initiation of a recovery should be instinctive (unless a full stall is being practiced intentionally from an altitude that allows recovery at least 1,500 feet above ground level [AGL] for single-engine airplanes and 3,000 feet AGL for multiengine airplanes). One indication of a stall is a mushy feeling in the flight controls and less control effect as the aircraft's speed is reduced. This reduction in control effectiveness is attributed in part to reduced airflow over the flight control surfaces. In fixed pitch propeller airplanes, a loss of revolutions per minute (RPM) may be evident when approaching a stall in power-on conditions. For both airplanes and gliders, a reduction in the sound of air flowing along the fuselage is usually evident. Just before the stall occurs, buffeting, uncontrollable pitching, or vibrations may begin. Many aircraft are equipped with stall warning devices that will alert the pilot 4 to 8 knots prior to the onset of a stall. Finally, kinesthesia (the sensing of changes in direction or speed of motion), when properly learned and developed, will warn the pilot of a decrease in speed or the beginning of a **mushing** of the aircraft. These preliminary indications serve as a warning to the pilot to increase airspeed by adding power, lowering the nose, and/or decreasing the angle of bank.

9.4 Types of Stalls

Stalls can be practiced both with and without power. Stalls should be practiced to familiarize the student with the aircraft's particular stall characteristics without putting the aircraft into a potentially dangerous condition. In multiengine airplanes, single-engine stalls must be avoided. Descriptions of some different types of stalls follow.

9.4.1 Power-off Stalls (also known as approach-to-landing stalls)

Power-off stalls are practiced to simulate normal approach-to-landing conditions and configuration. Many stall/spin accidents have occurred in these power-off situations, such as crossed control turns from base leg to final approach (resulting in a skidding or slipping turn); attempting to recover from a high sink rate on final approach by using only an increased pitch attitude; and improper airspeed control on final approach or in other segments of the traffic pattern.

9.4.2 Power-on Stalls (also known as departure stalls)

Power-on stalls are practiced to simulate takeoff and climbout conditions and configuration. Many stall/spin accidents have occurred during these phases of flight, particularly during go-arounds. A causal factor in such accidents has been the pilot's failure to maintain positive control due to a nose-high trim setting or premature flap retraction. Failure to maintain positive control during short field takeoffs has also been a causal accident factor.

9.4.3 Accelerated Stalls

Accelerated stalls can occur at higher-than-normal airspeeds due to abrupt and/or excessive control applications. These stalls may occur in steep turns, pullups, or other abrupt changes in flightpath. Accelerated stalls usually are more severe than unaccelerated stalls and are often unexpected because they occur at higher-than-normal airspeeds.

9.5 Stall Recovery

The key factor in recovering from a stall is regaining positive control of the aircraft by *reducing the AOA*. At the first indication of a stall, the aircraft AOA must be decreased to allow the wings to regain lift. Every aircraft in upright flight may require a different amount of forward pressure or relaxation of elevator back pressure to regain lift. It should be noted that too much forward pressure can hinder recovery by imposing a negative load on the wing. The next step in recovering from a stall is to smoothly apply maximum allowable power (if applicable) to increase the airspeed and to minimize the loss of altitude. Certain high performance airplanes may require only an increase in thrust and relaxation of the back pressure on the yoke to effect recovery. As airspeed increases and the recovery is completed, power should be adjusted to return the airplane to the desired flight condition. Straight-and-level flight should be established with full coordinated use of the controls. The airspeed indicator or tachometer, if installed, should never be allowed to reach high speed red lines at any time during a practice stall.

9.6 Secondary Stalls

If recovery from a stall is not made properly, a secondary stall or a spin may result. A **secondary stall** is caused by attempting to hasten the completion of a stall recovery before the aircraft has regained sufficient flying speed. When this stall occurs, appropriate forward pressure or the relaxation of back elevator pressure should again be performed just as in a normal stall recovery. When sufficient airspeed has been regained, the aircraft can then be returned to straight-and-level flight.

9.7 Intentional Spins

The *intentional spinning* of an airplane, for which the spin maneuver is not specifically approved, is NOT authorized by this handbook or by the Code of Federal Regulations. The official sources for determining if the spin maneuver IS APPROVED or NOT APPROVED for a specific airplane are:

- Type Certificate Data Sheets or the Aircraft Specifications.
- The limitation section of the FAA-approved AFM/POH. The limitation sections may provide additional specific requirements for spin authorization, such as limiting gross weight, CG range, and amount of fuel.
- On a placard located in clear view of the pilot in the airplane, NO ACROBATIC MANEUVERS INCLUDING SPINS APPROVED. In airplanes placarded against spins, there is no assurance that recovery from a fully developed spin is possible.

There are occurrences involving airplanes wherein spin restrictions are *intentionally* ignored by some pilots. Despite the installation of placards prohibiting intentional spins in these airplanes, a number of pilots, and some flight instructors, attempt to justify the maneuver, rationalizing that the spin restriction results merely because of a "technicality" in the airworthiness standards. Some pilots reason that the airplane was spin tested during its certification process and, therefore, no problem should result from demonstrating or practicing spins. However, those pilots overlook the fact that a normal category airplane certification only requires the airplane recover from a one-turn spin in whichever direction takes longer. This same test of controllability can also be used in certificating an airplane in the Utility category (14 CFR section 23.221 (b)).

The point is that 360° of rotation (one-turn spin) does not provide a stabilized spin. If the airplane's controllability has not been explored by the engineering test pilot beyond the certification requirements, prolonged spins (inadvertent or intentional) in that airplane place an operating pilot in an unexplored flight situation. Recovery may be difficult or impossible.